PRESENTATION - OF DATA IN SCIENCE

Publications, slides, posters, overhead projections, tape-slides, television

Principles and practices for authors and teachers

by

LINDA REYNOLDS

and

DOIG SIMMONDS

1984 **MARTINUS NIJHOFF PUBLISHERS**
a member of the KLUWER ACADEMIC PUBLISHERS GROUP
DORDRECHT / BOSTON / LANCASTER

Distributors

for the United States and Canada: Kluwer Academic Publishers, 190 Old Derby Street, Hingham, MA 02043, USA
for the UK and Ireland: Kluwer Academic Publishers, MTP Press Limited, Falcon House, Queen Square, Lancaster LA1 1RN, England
for all other countries: Kluwer Academic Publishers Group, Distribution Center, P.O. Box 322, 3300 AH Dordrecht, The Netherlands

Library of Congress Cataloging in Publication Data   CIP

Reynolds, Linda.
    Presentation of data in science.

    Bibliography: p.
    Includes index.
    1. Scientific illustration. 2. Communication in science.   I.   Simmonds, Doig.   II.   Title.
Q222.R48   507'.8   81-3976
ISBN 90-247-2398-1 (hard cover)   AACR2
ISBN 90-247-3054-6 (paperback)

First edition 1981
Second printing 1982
Third printing 1983
Fourth printing 1984
Fifth printing 1989

Copyright

PRESENTATION OF DATA IN SCIENCE

Contents

6. Slides   83

7. Posters   110

11. Materials, equipment and working comfort   136

12. Basic techniques   160

X

List of illustrations

Chapter 4

Chapter 5

Chapter 6

Chapter 7

Chapter 13

Notes on the illustrations

Figures: 1.1, 1.5, 1.6, 1.8, 2.1, 2.2, 2.3, 2.4, 2.5, 3.1, 3.2, 3.3, 3.4, 3.5, 3.6, 3.7, 3.8, 3.9, 4.1, 6.5, 6.6, 6.7

Supplied by Linda Reynolds, RCA, London

Figures: Introduction, 1.2, 1.3, 1.4, 1.7, 3.10, 4.2, 4.3, 4.4, 4.5, 4.6, 4.7, 4.8, 4.9, 4.10, 4.11, 4.12, 5.1, 5.2, 5.3, 5.4, 5.5, 5.6, 5.7, 6.1, 6.2, 6.3, 6.4, 6.8, 6.9, 6.10, 6.11, 7.1, 7.2, 7.3, 8.1, 8.2, 8.3, 8.4, 9.1, 10.1, 10.2, 11.2, 11.3, 11.4, 11.6, 12.1, 12.2, 12.3, 12.4, 12.5, 12.6, 12.7, 12.8, 12.9, 12.10, 12.11, 12.12, 12.13, 12.14, 12.15, 12.16, 12.17, 12.18, 13.1

Drawn by Doig Simmonds, RPMS, London

Figures: 5.8, 11.1, 11.5

Photographed by William Hinkes, RPMS, London

Foreword

'Do-it-yourself' is common advice in every walk of life. But how? If you own a home the problem is solved simply: pay a decorator to paint and paper a room while acting as his assistant. Watch how it is done and then try for yourself. The result may not be perfect but with practice it can be pretty good and you'll save money. Moreover it can be fun and something to be proud of. The house-owner becomes more adventurous and gradually collects equipment from the DIY shop, where advice from the owner and other customers is given freely and willingly. As a result the DIY scene has changed in twenty years to a multi-million pound business. We all do it and invite others to come and praise our work.

And what of medical and scientific illustration? Most authors of papers for publication don't even try. Yet the principles are the same: practice and precept. See how it is done and then do it.

From the Royal Postgraduate Medical School and Hammersmith Hospital there are about 700 publications every year: scientific advances, reviews of subjects, case reports. If, on average, each paper requires four drawn illustrations – tables, diagrams, graphs, charts – the total is considerable, and it would need four full-time illustrators to cope with that load alone. But there are almost the same number of lectures each year with an even greater demand for visual information. So what can be done? Firstly, learn the techniques of presenting data in the most easily assimilable form for the reader or listener. Secondly, do it yourself.

There are personal benefits of doing it yourself. The house-holder learns from his mistakes and, when anything goes wrong later, he will know where to look; but he also learns a great deal about his house – a kind of higher domestic education. So too with illustrating your own ideas. If your thoughts are not clear in your own mind, then how on earth can you express them on paper? In writing we have rules of language – grammar, spelling, punctuation – which a colleague will apply to the paper you ask him 'to look over' before sending it to an editor who will most certainly return your gem if it does not comply with accepted literacy. The editor may also reduce the number of illustrations, either because there are too many or because he cannot understand them, for his job is to protect his readers and boost his circulation. In the past, authors and lecturers asked the Medical Artist to make an illustration, an imperious demand which was difficult to refuse. As a confirmed lecture-goer (some prefer plays or opera) I have sat through dissertations where not only was there too much data on a slide but even some slides where the speaker never looked up from his script to explain (or, if he did, it was to tell the audience to 'ignore the first three columns'). This is not good enough. Already, editors have become tough, by suggesting that tables of data should be made available from the author if required, so that the published article is slimmed down to readable proportions. For the guest lecturer, hosts have reduced the time for exposition in the hope that over-crowed slide No. 100 will not be shown. TV does it better because viewing figures tell a producer how others feel about his presentation of data.

Why make a complex matter more difficult? Why make a simple matter complex? The whole idea of writing and speaking is to transmit information to another person: if visual display does it better than words, then do so, but as in language make the message clear and concise. The first rule in TV is 'if you can show it, don't speak it'. The same in scientific publications.

This book tells you how to do it well. It is largely based on a do-it-yourself studio started at the Royal Postgraduate Medical School five years ago by Doig Simmonds. In his words, it even allows the handicapped (those who can't draw a straight line with a ruler) to produce work of high quality. Does it work? Well, the total number of illustrations performed by these so-called amateurs has risen by 30% each year. And like a well-known advertisement, you can't tell the difference. If you do have to produce your own illustrations, why spoil a good paper by poor pictures? There are plenty of books to help the DIY householder, but none for the lecturer and writer of science and medicine. *The Presentation of Data in Science* is, as far as I know, the only manual which caters specifically for such needs: it tells you how to do it and explains why.

I have pleasure in recommending this book to all who aspire to writing a paper for publication and to those who are forced to lecture on their favourite subject. As Seneca said 2,000 years ago: 'there is no one less fortunate than he whom adversity neglects: he has had no chance to prove himself'. So why not read this book and try?

Valentines Day 1980   JAMES CALNAN, FRCS, FRCP
Professor of Plastic and Reconstructive Surgery
in the University of London
at the Royal Postgraduate Medical School.

Acknowledgements

I

The chapters in this book which deal with the more theoretical aspects of data presentation have grown from a paper originally written for inclusion in 'Data in medicine: Collection, processing and presentation' edited by Robert Reneman and Jan Strackee. I am very grateful to Herbert Spencer (Professor of Graphic Arts at the Royal College of Art) and Robert Reneman (Professor of Physiology at the University of Limburg) for their advice and help in preparing the original paper. I should also like to thank Professor Reneman and Mr. Boudewijn Commandeur of Martinus Nijhoff for encouraging me to broaden the scope of that original paper and to present it in book form, together with Doig Simmonds' practical advice on standards, materials and techniques for the production of artwork.

London 1980   Linda Reynolds

II

I owe more than can be expressed in a few words to my teacher, Sylvia Treadgold, who was Senior Medical Artist at Guy's Hospital when I went there as one of her students in 1952. Sylvia taught that a 'successful' drawing (one that properly fulfills its function) was largely the result of common sense combined with a knowledge of processes, plus the ability to draw. The most important consideration being that all the materials and processes used in the graphic arts have their limitations, and that the good designer keeps within the confines imposed by them.

During the 50's and 60's, Charles Engel and Francis Speed were both very accomplished medical photographers and from them I learnt that most scientific illustrations have to go through photographic procedures which in themselves imposed yet more constraints. An interest in photography is of considerable help to any graphic artist and I owe a lot to their teaching.

Professor R.G. Hendrickse of the Liverpool School of Tropical Medicine asked me to provide a three day course in Do-It-Yourself basic statistical illustration for his postgraduate tropical medicine students and this is now done regularly each year. I am very grateful for this opportunity to learn more about communicating ideas to other people.

Several manufacturers have given me considerable help and advice and some of them have developed new products especially designed to help illustrators, both professional and amateur to achieve better results more quickly.

Gary Gillott of Printaids Ltd. was particularly helpful as a consultant on dry transfer symbols and gave me very useful instruction on the technology involved. Colin Cheesman, Head of Graphics, BBC, London and his team gave me valuable advice, and I have had a great deal of encouragement and practical help from Sheila Carey, who is one of the Directors of Butterworths Medical Publishers.

The Institute of Medical and Biological Illustration have done so much in this country to encourage the application of standards to the profession of medical illustration; it was while working on their publication 'Charts and Graphs' that I realised the need for a companion volume on how-to-do-it methods and, to some extent, this book is a response to the need. I am very grateful to IMBI for the support they have given and for allowing me to draw heavily on the information in 'Charts and Graphs'.

My greatest debt is to the large number of Research and Teaching staff, secretaries and others, at the Royal Postgraduate Medical School who have responded to my attempts at standardising the principles of drawing for publication and slides. If there has been anything approaching a successful outcome of this joint venture, it is very much due to their understanding, forebearance and willingness to take part in the experience.

Last but not least, I have had consistent help and co-operation from Glyn Williams and all other members of the Department of Medical Illustration at RPMS.

London 1980   Doig Simmonds

Introduction

A vitally important part of any research activity is the dissemination of the results to other research workers, to students, and to those in a position to implement the findings in a practical way. Unless this dissemination process is carried out efficiently, the results will not become widely known or used and much of the time and money expended on the research will have been effectively wasted.

The dissemination of research results can take place in a number of ways, but in most cases some form of visual presentation of all or part of the information is necessary. Printed materials in the form of reports, journal articles and books are perhaps the most commonly used medium for publicising research, and here the presentation is entirely visual. At conferences and in other lecture situations, visual aids such as slides and overhead projector transparencies are used as an essential complement to the spoken word. More recently, posters have become popular as a means of presenting research results. In this case, a visual summary is presented under circumstances where informal discussions with the author are possible. In teaching situations, tape-slide programmes and closed circuit television are useful visual aids, and in both cases visual information is used in conjunction with the spoken word.

The effective communication of information in visual form, whether it be text, tables, graphs, charts or diagrams, requires an understanding of those factors which determine the 'legibility', 'readability' and 'comprehensibility', of the information being presented. By legibility we mean: can the data be clearly seen and easily read? By readability we mean: is the information set out in a logical way so that its structure is clear and it can be easily scanned? By comprehensibility we mean: does the data make sense to the audience for whom it is intended? Is the presentation appropriate for their previous knowledge, their present information needs and their information processing capacities? For example, a series of slides on a technical subject for a lay audience is likely to differ somewhat in terms of content and design from a similar series intended for an audience already familiar with the subject. What is comprehensible to the latter may be nonsense to the former.

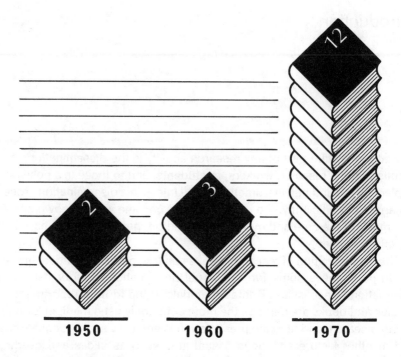

1950   1960   1970

The growth of medical literature as reflected in the number of volumes of Index Medicus published each year.

Professional illustration sources available to the 'life sciences' have not grown at the same rate, and many illustrations are now undertaken by non-professionals. This book is intended as a guide for the author/illustrator.

The comprehensibility of data is the prime concern of the teacher and illustrators have to presume that the teacher understands the processing capacities of his audience. On the other hand, legibility and readability are the prime concern of the illustrator and the teacher may have to modify his data to suit the limitations imposed by the medium selected to impart a given piece of information. It is vital to remember that *data presented in one form are not necessarily suitable for another*. For example, the design of a printed illustration can be a very different matter from the design of a slide giving similar information, because each medium has its own particular characteristics, its advantages and its limitations. Thus, in the case of printed materials it is possible to present a considerable amount of detail in the form of text, tables and illustrations, because the user is able to study the information at his own pace. The content of slides and OHP transparencies, however, must be restricted to relatively simple illustrations and very small amounts of text if the audience is to be able to assimilate the information in the time allowed by the presenter. Posters lend themselves to the presentation of more detail than would be appropriate for slides or OHP transparancies, but they are not suitable for the level of detail which would be expected in a written paper. In tape-slide programmes it is usually possible to give more information than in lecture slides because the reader is self-paced, but television displays need to be kept very simple because the user has no control over the speed of presentation. *The design of text, tables and illustrations must therefore take into account the characteristics of the medium concerned.*

The design of scientific texts, tables and illustrations is, therefore, by no means a simple matter, yet it is very important that the presentation should be of a high standard. The quality of the presentation will not only affect the ease and speed with which potential users of research results are able to read, understand and remember the information given; it will also create an image which will often be extended to the quality of the research itself in the minds of the audience. Badly designed and carelessly executed visuals can not only create confusion and lead to misunderstandings, they can also give an overall impression of sloppiness which may be totally undeserved. *It is therefore essential to adopt the same professional approach to the presentation of the results as to the research work.*

Some of the larger research and teaching establishments have art departments manned by specialists who are trained to produce visual materials of a high standard. All too often, however, the time taken for work to be processed may be unacceptably long, especially where artwork is required at short notice. Many smaller establishments have no art department at all. The situation often arises, therefore, where the researcher or teacher is forced to design and prepare his own artwork. Few workers in the life sciences have any formal training in visual communication or graphic design and the need to produce visual materials is often seen as an almost insuperable problem. Difficulty in producing illustrations is often given as a reason for delayed publication or reluctance to give a lecture.

The preparation of illustrations and other visual materials need not be the obstacle which it oftenseems. The design and execution of visual materials which will be adequate for publication or teaching purposes is a skill which can be learned by the amateur, provided that he takes the trouble to become aware of certain basic principles, uses good quality materials and equipment, and is prepared to learn from his own experience. At the Royal Postgraduate Medical School in London in 1977, some 6000 illustrations were produced to publishable standards by amateurs, using materials and equipment provided by the art department. Some 1400 persons are recorded as having used these facilities over the year. There is no doubt, therefore, that the preparation of adequate illustrations is within the capabilities of most scientists if they are willing to try.

The aim of this book is to provide the amateur with an awareness of basic principles and practices in the preparation of visual materials. We begin in Chapter 1 with a review of the factors which affect the legibility of individual characters, words and continuous text. This is followed by a discussion of the factors which affect the readability or comprehensibility of text (Chapter 2), tables (Chapter 3), and graphs charts and diagrams (Chapter 4). We then go on to look at the characteristics of five different presentation media, and at the graphic standards which must be observed if information presented in these media is to be legible (Chapters 5-10). These theoretical considerations are followed by practical advice on essential materials and equipment (Chapter 11) and on basic techniques in the preparation of the artwork (Chapter 12). Finally, we discuss working method, from the planning stage of illustration through to completion of the final artwork (Chapter 13).

Preface to second printing

Throughout this book the authors have mentioned the use of the IBM Executive typewriter with the Directory type face. This machine is now obsolete and the alternatives do not have such a large or clear letter form. This can make for difficulties in art-work production since the size of drawing may have to be somewhat smaller than the 13 cm × 20 cm formats specified in the book. However, the design principles remain the same, whatever typeface, size of style is employed. The rule is: measure the capital letter height of the available lettering, multiply this by forty and the result is the optimum measurement for the long side of the artwork.

In this second printing a number of minor errors from the first printing have been corrected.

1. The legibility of type

1.1. Introduction

The word 'legibility' is used here to mean the ease and speed of recognition of individual letters or numerals, and of words either singly or in the form of continuous text.

The first recorded experiments on legibility were conducted over 150 years ago, though it was not until this century that typographic designers and psychologists began to investigate systematically the effects of typographic variables on reading performance. Until relatively recently legibility research has been concerned mainly with the design of printers' typefaces and the setting of continuous text, but the results of many of these studies have important implications for the presentation of information in other media besides print.

In more recent years legibility studies have been undertaken which relate specifically to forms of lettering other than printers' type, and to media other than the printed page. The following summary of legibility research relates to studies on printed materials which are also relevant to other situations; studies dealing specifically with legibility in the various media discussed in this book will be mentioned in the appropriate chapter.

The characteristics of the image itself are not, of course, the only factors influencing legibility. The viewing distance, viewing angle and the level of ambient illumination are also important factors, to name but a few. The importance of these and other factors depends to some extent on the medium in question and they will therefore be dealt with in later chapters.

1.2. Investigating legibility

1.2.1. The reading process

A knowledge of the perceptual processes involved in reading is essential for an understanding of the way in which typographic factors affect reading performance, or legibility. A number of early studies were concerned with the way in which words are recognised and these have been summarized by Spencer in 1968. As long ago as 1885, Cattell demonstrated that in a normal reading situation each word is recognised as a whole rather than as a series of individual letters, and Erdmann and Dodge concluded in 1898 that it is the length and characteristic shape of a word that are important for recognition. In 1905 Javal showed that the upper half of a word is more easily recognised than the lower half (Figure 1.1), and in 1903 Messmer suggested that letters

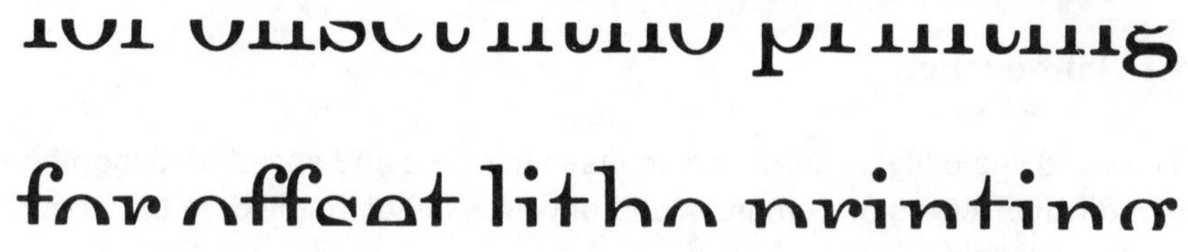

for offset litho printing

for offset litho printing

Figure 1.1   The upper half of a word is more easily recognised than the lower half.

with ascenders contribute most to word recognition. Similarly, in 1893 Goldscheider and Muller found that consonants, with their ascenders and descenders, contribute to characteristic word shapes more than vowels. Tinker, who spent many years investigating legibility, has stressed that it is the overall structure of the word – the internal pattern as well as the outline – that is important for recognition. It has been demonstrated recently, however, that rearrangement of the middle letters of words has very little effect on word recognition (Rawlinson 1975). This suggests that accepted theories about the way in which words are recognised are very much oversimplified.

Eye movement studies have contributed much to our knowledge of the reading process. The eyes move along a line of print in a series of jerks, or saccadic movements, during which no clear vision is possible. At the end of each movement there is a fixation pause lasting approximately 200 to 250 milliseconds, during which perception occurs. On average, 94 per cent of reading time is devoted to fixation pauses and the remainder to interfixation movements. Sometimes the eyes make a backward movement or regression to re-examine material not clearly perceived or understood and at the end of each line they make a return sweep to the beginning of the next. The frequency and duration of fixations and regressions are increased by poor typography and texts which are too complex in their content.

At a normal reading distance of 30 to 35 cm, only about four letters of normal size print fall within the zone of maximum clearness or foveal vision. The field of peripheral vision usually encompasses about twelve or fifteen letters on either side of the fixation point. During a single fixation it may therefore be possible to read up to thirty letters in word form, as opposed to only three or four unrelated letters. Short exposure experiments have shown, however, that most adults can perceive in their field of vision about three or four words in 1/50th of a second. Allowing for the time occupied by movements between fixations, this means that the eyes are capable of reading ten times more quickly than they normally do, which suggests that the reader is ultimately limited by his rate of comprehension.

1.2.2. Methods of research

There are numerous definitions of the term 'legibility'; it is used in relation to studies ranging from the visibility and perceptibility of individual characters and words to the ease and speed of reading of continuous text. A precise definition of legibility can only be given in relation to the technique used for measuring it. A number of methods have been used, and these vary in their validity, i.e. the extent to which they measure that which they are intended to measure, and in their reliability. Different techniques are appropriate for different reading situations and for different types of reading materials, and results are rarely relevant to conditions other than those tested. Tinker (1963) and Zachrisson (1965) have published detailed descriptions of the principal techniques, and some of them are summarised below.

Speed of perception. Speed of perception is measured by means of a tachistoscope, which is used to give very short exposures of individual characters, words or phrases.This method is useful for determining the relative legibility of different characters and of different character designs, but results have little relevance to the reading of continuous text.

Perceptibility at a distance. Perceptibility is often measured in terms of the maximum distance from the eye at which printed characters can be recognised. This method is appropriate when used to assess the relative legibility of characters which are intended to be read at a distance as on slides and posters, but can be misleading if applied to continuous text intended to be read under normal reading conditions.

Perceptibility in peripheral vision. This is assessed by measurement of the horizontal distance from the fixation point at which a printed character can be recognised. It has been used to compare the relative legibility of single characters and to compare the legibility of black letters on white with white on black.

Visibility. Visibility is usually measured by means of the Luckiesh-Moss Visibility Meter. This piece of equipment allows the contrast between image and background to be varied by means of filters. It has been used to determine the threshold contrast, and hence the relative legibility, of different type

3

faces, sizes and weights, and to examine the effects of variation in contrast between image and background.

Blink rate. Reflex blink rate has been used as a measure of legibility by Luckiesh and Moss, but the validity of the method is questionable. The assumption is that the blink rate will increase as legibility decreases. The same authors also advocated heart rate as a measure of legibility, but this method is extremely suspect.

Visual fatigue. This has been suggested as a criterion of legibility, but no satisfactory objective method of measurement has been devised. Subjective assessments of fatigue are subject to modification by a great many factors which may be totally unrelated to the experimental situation.

Eye movement studies. Eye movement records provide a very useful insight into the effects of typographic factors on reading performance. They indicate whether a slower rate of reading resulting from poor typography is associated with more frequent fixations, longer fixations, more regressions, or some combination of these. Total perception time, i.e. the time spent in fixation pauses, and fixation frequency have been found to be valid and highly reliable measures of legibility. The duration and frequency of regressions have been found to be less reliable.

Rate of work. Rate of work is by far the most satisfactory measure of legibility. This can be assessed either in terms of the time required to complete a given reading task, or in terms of the amount of work done within a set period. A number of methods of measurement have been used, including speed of reading continuous text (with or without a comprehension check), speed of scanning text in search of target words, and "look-up" tasks of various kinds on structured information such as indexes and directories. Rate of work is undoubtedly one of the most valid measures of legibility because the task can be directly related to the normal use of the material under test. Most of the methods used are also highly reliable when carefully administered. There has been some dispute as to the time periods necessary for these methods. Tinker (1963) maintains that differences which are significant after 1 minute 45 seconds will continue to be so after 10 minutes, but he accepts that significant differences may emerge after 10 minutes which would not be apparent during a short reading period.

Readers' preferences. Readers' preferences cannot be relied upon as an accurate indication of the relative legibility of printed materials. Tinker carried out an experiment in which he compared measurements of visibility, perceptibility at a distance, reading speed and judged legibility for ten different typefaces. He found that judged legibility corresponded more closely with visibility and perceptibility than with speed of reading. Tinker and Paterson studied the relation between judgements of 'pleasingness' and judgements

4

of legibility for a number of typographical variables. They found that in general the agreement between the two sets of judgements was high. Tinker concluded from these studies that readers prefer those typefaces or typographical arrangements which they can read most easily (Tinker, 1963). Burt, however, put forward the argument that readers are able to read most easily those typefaces which they find most pleasing (Burt, 1959).

1.3. Type forms

1.3.1. The basic anatomy of type

Until relatively recently the majority of printed materials were set in metal type, and many of the words now used to describe the various dimensions and characteristics of lettering of all kinds were originally coined in relation to metal type. The meaning of some of these terms is illustrated in Figure 1.2.

With the advent of phototypesetting and other forms of lettering which do not involve the use of metal type, some of these terms have lost their literal meaning.

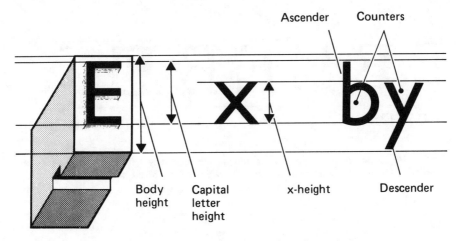

Figure 1.2   The anatomy of type. Note: x, b, y, etc. are 'lower case' letters.

1.3.2. Typeface

Much early research was concerned with establishing the relative legibility of different typefaces for continuous text. Different typefaces of the same body size may not have the same x-height or capital height however (see Figure 1.3), and many of these studies compared types of the same body size rather than of the same visual size or x-height. The effects of letter form could not,

SANS SERIF FAMILY

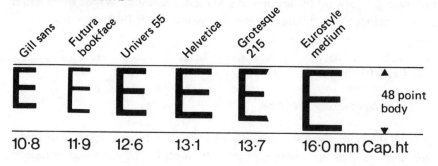

Gill sans	Futura bookface	Univers 55	Helvetica	Grotesque 215	Eurostyle medium	
10·8	11·9	12·6	13·1	13·7	16·0 mm Cap.ht	48 point body

SERIF FAMILY

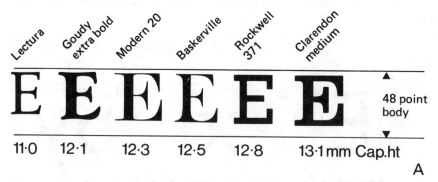

Lectura	Goudy extra bold	Modern 20	Baskerville	Rockwell 371	Clarendon medium	
11·0	12·1	12·3	12·5	12·8	13·1 mm Cap.ht	48 point body

A

Figure 1.3A Capital letter height as measured in millimetres may differ with each typeface, in spite of the fact that the point measurement (or body size) may be the same.

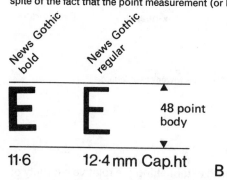

News Gothic bold	News Gothic regular	
11·6	12·4 mm Cap.ht	48 point body

B

B Even in the *same* typeface, capital letter height can vary between styles.

6

abcde   Regular or Roman

abcde   Italic

abcde   Light

abcde   Condensed

abcde   Extended or Expanded

Figure 1.3C   All of the above examples are of the same point size (48 point Helvetica). Note how the various styles occupy different amounts of horizontal space.

therefore, be separated from the effects of letter size. It would seem, however, that there is very little difference in the legibility of typefaces in common use when well printed, and that familiarity and aesthetic preference have much to do with any observed differences in reading speed.

The relative legibility of seriffed and sans serif faces has been a particularly controversial issue. Some examples of these two type families are shown in Figure 1.4. It has been argued that serifs contribute to the individuality of the letters, that they are responsible for the coherence of letters into easily recognisable word shapes, and that they guide the eye along each line. None of these claims has been conclusively proved, however, and it would seem that any differences in legibility in favour of seriffed faces may be due to familiarity rather than to any intrinsic superiority. On the other hand, there is evidence to suggest that sans serif faces are the more legible for children and

Size	Family	Style	Name
Point	Serif	Light	Univers
Millimetre	Sans serif	Regular	Helvetica
	Block serif	Medium	Rockwell
	Brush	Bold	News Gothic
	Display	Italic	etc.
	etc.	Extended	
		etc.	

Figure 1.4A   Type and its categories simplified.

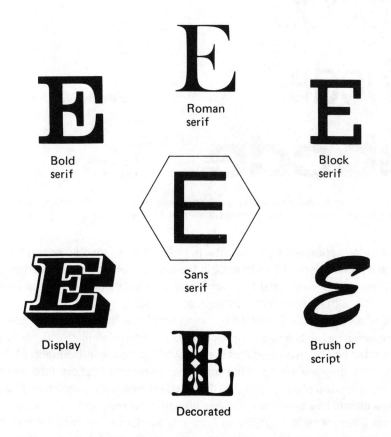

Bold
serif

Roman
serif

Block
serif

Sans
serif

Display

Decorated

Brush or
script

Figure 1.4B   Families of type. These are just a few examples of the available families of type. The sans serif family is the most effective for projection purposes (slides, OHP, etc.).

poor vision readers, and they are widely accepted as being suitable for display purposes because of their simple form.

The choice of typefaces is likely to influence considerably the legibility of any paper or microfilm copies of the information. Recent research has shown that some typefaces are better able to withstand extremes of image degradation than others (Spencer et al., 1977b). The typefaces tested are shown in Figure 1.5. The legibility of Baskerville, which has relatively thin strokes and a

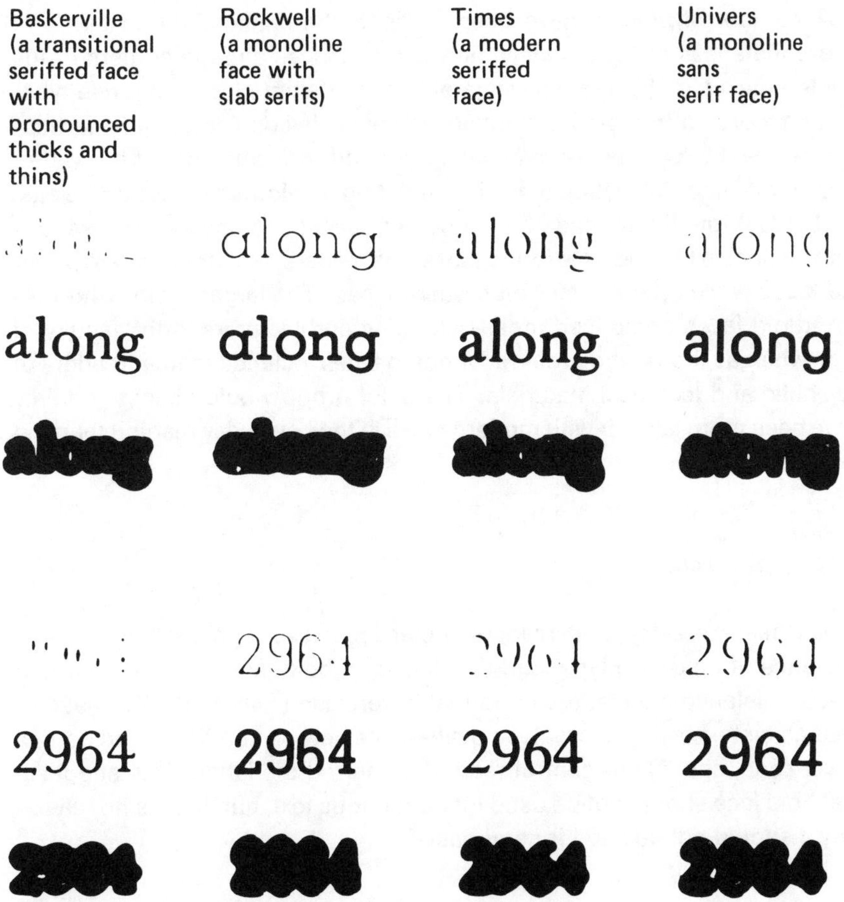
Baskerville (a transitional seriffed face with pronounced thicks and thins)
Rockwell (a monoline face with slab serifs)
Times (a modern seriffed face)
Univers (a monoline sans serif face)
along
along
along
along
2964
2964
2964
2964

Figure 1.5 Enlargements of 9 point type which has been photographically thinned-down and thickened-up; in each case a 'normal' image is shown for comparison.

Differences in the legibility of the typefaces are largely accounted for by differences in certain design characteristics such as x-height, normal stroke thickness, variation in stroke width within characters, the size of counters, the presence or absence of heavy serifs, and the spacing between characters.

small x-height in relation to its body size, was markedly impaired by any 'thinning-down' of the image. Rockwell, on the other hand, was severely affected by 'thickening-up' because of its relatively thick strokes and heavy serifs. Times and Univers were less affected by these extremes of degradation. This suggests that where thinning-down of the image is likely it is preferable to use typefaces which have a relatively large x-height and not too great a variation in stroke width. Where thickening-up is likely, it is best to choose a typeface which has large open counters and a generous set-width to x-height ratio (i.e. relatively generous letter spacing).

A number of studies have been made of the relationship, in terms of appropriateness or congeniality, between typefaces and the content of the printed message. Several investigators have attempted to categorise typefaces according to their 'atmosphere value'. This consideration is particularly relevant to persuasive material such as advertisements, but has rather less significance in relation to the presentation of information. Nevertheless, Burt (1959) found that readers of serious publications do have preferences with respect to typeface. He found that some readers had a tendency to prefer old faces while others preferred modern faces. The larger group who preferred old faces comprised mostly students and lecturers in the faculty of arts, whereas those who preferred modern faces included regular readers of scientific and technical materials. The latter group would almost certainly have been more familiar with modern faces in their everyday reading than the former group.

1.3.3. Type weight

It has been reported that readers prefer typefaces 'approaching the appearance of bold face type' and that letters in bold face are perceived at a greater distance than letters in normal lower case (Tinker, 1963). Speed of reading tests, however, revealed no difference between bold face and normal lower case, and 70 per cent of readers preferred the latter. This suggests that bold face should not be used for continuous text, but there is no reason why it should not be used for emphasis.

1.3.4. Italics

Experiments have shown that the use of italics for continuous text retards reading, especially in small typesizes and under poor illumination. Further, readers apparently do not like italics; 95 per cent prefer 'roman' lower case

(Tinker, 1963). It is unlikely, however, that the occasional use of italics for emphasis will significantly impair legibility.

1.3.5. Capitals versus lower case

All-capital printing has been shown to markedly reduce the speed of reading of continuous text. Reductions of 13.9% over a 20 minute reading period have been recorded (Tinker, 1963). This is partly because words in lower case have more distinctive shapes than words in capitals and may therefore be easier to recognise, and partly because text set entirely in capitals occupies about 40 to 45% more space than text in lower case of the same body size. Eye movement studies have shown that all-capital printing increases the number of fixation pauses and that, because the text occupies a larger area, the number of words perceived at each fixation is reduced. These findings are very significant with respect to upper case computer printout and computer output on microfilm. As a general rule, sections of upper-case-only com-puter printout should not be reproduced directly for inclusion in a printed document. The data should be typeset or typed.

Only under exceptional circumstances are capitals likely to be more legible than lower case. This may happen, for example, where very small type sizes approaching the threshold of visibility are used. Experiments have shown that capitals may then be more easily discriminated than corresponding lower case letters because of their greater size. For the same reason they may be more legible than lower case letters when the limit for distance vision is approached.

1.3.6. Numerals

The legibility of numerals is particularly important because each individual character must be identifiable and contextual cues are usually weak or absent. Tinker carried out tests on Modern and Old Style numerals (see Figure 1.6). He found that Old Style numerals, which vary in height and alignment, were more easily recognised at a distance than Modern numerals which are all the same height, but in normal reading situations the two kinds of numeral were read equally fast and with equal accuracy.

Arabic numerals are read significantly faster and much more accurately than Roman numerals; this is because the former are so much more familiar. The differences are so marked, particularly for numbers greater than ten, that there is little justification for using Roman numerals in most cases (Tinker, 1963).

Modern: 7 4 1 6 9 0 2 3 8 5
Old Style: 7 4 6 0 1 9 3 5 2 8

Figure 1.6A   Tinker ranked isolated numerals in this order of legibility.

Modern: 7 1 4 0 2 9 8 5 6 3
Old Style: 8 7 6 1 9 4 0 5 3 2

B. When the numerals were arranged in groups this was the order.

1.4. Type size

Type size is conventionally expressed in terms of the *body* size measured in points (see Figure 1.2). There are 72 points to the inch, and one point is approximately 0.38 mm. From the designer's point of view, however, the capital letter height is often a more practical measure than the body size.

It is possibly true to say that type size has more effect on legibility than any other single typographic factor, and much research has been devoted to the relative legibility of different type sizes for printed materials. The most reliable investigations suggest that the more commonly used sizes between 9 and 12 point are of about equal legibility at normal reading distance (Tinker, 1963). The optimum size is likely to be 10 or 11 point, depending on the x-height of the typeface in relation to its body size. Assuming that the reading distance remains constant, the legibility of smaller sizes such as 6 point will be impaired by difficulties encountered in discriminating letters and recognising words. Larger sizes such as 14 point will reduce the number of characters which fall within the area of maximum visual acuity on the retina, and as a result, fewer words will be perceived at each fixation. This will be counter-balanced to some extent, however, by the fact that larger characters will be more easily recognisable in peripheral vision.

Optimum sizes for larger displays such as slides and posters have not been so thoroughly researched, but it is clear from the available results that the size of the image on the retina is the significant factor. As viewing distance is increased, type size must also be increased so that the size of the retinal image remains more or less constant.

1.5. Line length

The line length or 'measure' of a piece of text is often expressed in picas. One pica is equal to 12 points, or approximately 4 mm.

Line length may be varied within broad limits without diminishing legibility, but research has shown that very short lines prevent maximal use of peripheral vision and thus increase the number and duration of fixations and hence total perception time. Very long lines, on the other hand, cause difficulty in locating the beginning of each successive line, and the number of regressions after the backsweep to the beginning of each line is greatly increased (Tinker, 1963). These effects are especially noticeable in the very short lines of some newspapers, and in the very long lines of up to 132 characters which are typical of many kinds of computer output. The effect of short lines is particularly severe with large type sizes, as is the effect of long lines with small type sizes. Line length cannot be determined independently of type size therefore.

From the point of view of legibility, it is more meaningful to think of line lengths in terms of the number of characters and spaces, rather than in terms of a linear measurement. The optimal line length for conventionally printed materials seems to be one which accomodates about ten to twelve words, or sixty to seventy characters. Somewhat shorter lines of about forty to fifty characters may be more appropriate for larger displays however. The reason for this stems from the fact that the character size of any large display must be chosen with the farthest reader in mind. For those closer to the display, fewer characters will fall within the area of maximum visual acuity on the retina and more fixations per line will therefore be necessary. This may give rise to an uncomfortable awareness of the physical distance which must be scanned in reading each line.

1.6. Line spacing

The space between lines of metal type is created by the type body, which extends slightly beyond the printing surface of the character (see Figure 1.2). If more space is required between lines, this can be achieved by casting the letters on a larger type body. Thus '9 on 10 point' would refer to a 9 point letter cast on a 10 point body. Alternatively, narrow strips of lead may be inserted between lines of 9 point type to create the same effect. In this case the type would be referred to as '9 point, 1 point leaded'. If no leading is used, the type is said to be 'set solid'. The term 'leading' is widely used to refer to the amount of space between lines of type, though in the case of phototypeset materials it has no literal meaning.

It would appear from the experimental evidence available that the use of leading improves the legibility of most sizes of type, but that small sizes benefit most. The additional space is helpful in locating the beginning of successive lines, and the effect is enhanced for smaller sizes because of the greater number of characters in a line of any given length. If, however, the type is too small to be discriminated easily – or too large to allow the maximum number of characters to be perceived in each fixation – then leading will have little effect.

Leading appreciably modifies the effect of line length on legibility, tending to increase the maximum acceptable length for a given type size (Tinker, 1963). This is particularly true for smaller sizes. Experiments suggest that the optimum amount of leading for 8 or 10 point type is 1 or 2 points (Spencer, 1969). The results of these studies show clearly that line length and leading interact in a complex way and must therefore be selected in relation to one another.

With respect to typewritten materials, if a line length of maximum 60 characters is used, then the minimum line spacing available on the type-writer is usually adequate for most purposes. (Proofs which require space for corrections are, of course, a different matter.) More generous spacing will tend to fragment the information over a larger number of pages and may make it difficult for the reader to grasp the overall structure of the text (see Chapter 2).

1.7. Letter and word spacing

The space between adjacent letters should be sufficient to separate them, but not so great that they cease to hold together as words.

In the case of metal type, the amount of space between letters is deter-mined by the width of the type body in relation to the width of the character itself. The width of the type body will vary according to the width of the character, so that w and m, for example, will be cast on a wider body than i or 1. This is known as proportional spacing because the amount of space occupied by each character is proportional to its width. The average body width for all the letters in an alphabet is known as the 'set width'. This can be varied, and a larger set width will effectively open out the character spacing. Each character is positioned on its type body in such a way that when several characters are adjacent the spaces to the left and right of each of them add up to give an aesthetically pleasing and regular appearance. This apparent regularity of spacing is an ideal which should be sought after in all forms of lettering.

Relatively little experimental work has been carried out on the effects of letter spacing on legibility, but those results which are available suggest that small increases or decreases in 'normal' letter spacing have no appreciable effect on legibility (Spencer and Shaw, 1971). With phototypesetting, however, greater variation in spacing is possible because the physical limitation of the metal type body no longer exists. This means that characters can be – and sometimes are – set in such a way that they are actually touching. This can cause confusion in letter recognition, and very close spacing is certainly not advisable if the type image is likely to be thickened-up in any subsequent reproduction processes. Close spacing is also inadvisable for any form of negative image (see Section 5.2.9).

In the case of typewritten text from a standard typewriter, each character occupies the same amount of space regardless of its width, and this is known as monospacing. Typewritten text tends to have a much more uneven texture than typeset text as a result of the space created around narrow characters such as i and 1. Although many typographers find this aesthetically displeasing, it has not yet been shown conclusively that monospaced type is significantly less legible than proportionally spaced type when all other conditions are equal.

Word spacing should be such that it is sufficient to separate adjacent words, but not so great that vertical white 'rivers' are created on the page. In typewritten materials, word spacing will usually be one character space.

1.8. Justified versus unjustified setting

In justified setting, the type on each line extends exactly to the right hand margin as shown in Figure 1.7. This is achieved by varying the space between letters and words from line to line, and by hyphenating words at the ends of lines where necessary. In unjustified setting the spacing between letters and words is held constant, and each line is filled to the nearest whole word. Hyphenation is usually kept to a minimum, or not used at all.

Justification is quite a complex operation which is confined to the more sophisticated typesetting systems and typewriters. On most typewriters and many forms of computer output, justification is not possible. In recent years, unjustified setting has begun to become more popular for typeset materials because it offers advantages in terms of convenience and economy, and it is now used for a wide variety of printed materials including books, catalogues and reference works, magazines and newspapers.

At present there is no evidence to suggest that unjustified setting is any less legible than justified setting, and under certain circumstances the reverse may well be true. Experiments by several researchers have shown no significant differences in reading time or eye movement patterns for justified and unjustified text with moderate line lengths (Spencer, 1969). There is,

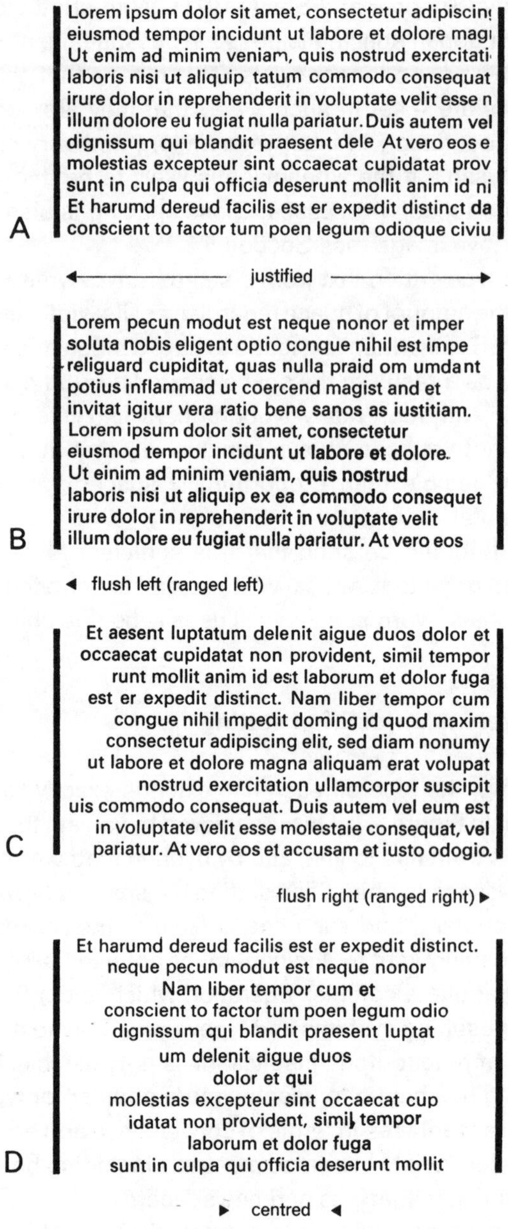

Lorem ipsum dolor sit amet, consectetur adipiscin
eiusmod tempor incidunt ut labore et dolore mag
Ut enim ad minim veniam, quis nostrud exercitati
laboris nisi ut aliquip tatum commodo consequat
irure dolor in reprehenderit in voluptate velit esse n
illum dolore eu fugiat nulla pariatur. Duis autem vel
dignissum qui blandit praesent dele At vero eos e
molestias excepteur sint occaecat cupidatat prov
sunt in culpa qui officia deserunt mollit anim id ni
Et harumd dereud facilis est er expedit distinct da
A conscient to factor tum poen legum odioque civiu

◄——————— justified ———————►

Lorem pecun modut est neque nonor et imper
soluta nobis eligent optio congue nihil est impe
religuard cupiditat, quas nulla praid om umdant
potius inflammad ut coercend magist and et
invitat igitur vera ratio bene sanos as iustitiam.
Lorem ipsum dolor sit amet, consectetur
eiusmod tempor incidunt ut labore et dolore.
Ut enim ad minim veniam, quis nostrud
laboris nisi ut aliquip ex ea commodo consequet
irure dolor in reprehenderit in voluptate velit
B illum dolore eu fugiat nulla pariatur. At vero eos

◄ flush left (ranged left)

Et aesent luptatum delenit aigue duos dolor et
occaecat cupidatat non provident, simil tempor
runt mollit anim id est laborum et dolor fuga
est er expedit distinct. Nam liber tempor cum
congue nihil impedit doming id quod maxim
consectetur adipiscing elit, sed diam nonumy
ut labore et dolore magna aliquam erat volupat
nostrud exercitation ullamcorpor suscipit
uis commodo consequat. Duis autem vel eum est
in voluptate velit esse molestaie consequat, vel
C pariatur. At vero eos et accusam et iusto odogio.

flush right (ranged right) ►

Et harumd dereud facilis est er expedit distinct.
neque pecun modut est neque nonor
Nam liber tempor cum et
conscient to factor tum poen legum odio
dignissum qui blandit praesent luptat
um delenit aigue duos
dolor et qui
molestias excepteur sint occaecat cup
idatat non provident, simil tempor
laborum et dolor fuga
D sunt in culpa qui officia deserunt mollit

► centred ◄

Figure 1.7 The alignment of type. The flush left version, *B*, is the most economical to typeset and the easiest to read. The centred version, *D*, is the most expensive to set and the slowest to read.

however, evidence to suggest that less skilled readers find unjustified text easier to read (Powers, 1962). Gregory and Poulton (1970) found that this was particularly true for very short lines of seven words, and they concluded that less skilled readers were confused by the uneven spacing and hyphenation associated with justification. The very obviously uneven spacing and 'rivers' created by the justification of short lines can also be very displeasing aesthetically.

Thus it would seem that the use of unjustified setting for continuous text may offer practical advantages with no disadvantages in terms of legibility. In the case of displayed text with very short lines, unjustified setting is likely to be preferable on grounds of aesthetics and legibility.

1.9. Margins

Margins conventionally used in books reduce the printed area to approximately half the total page area. Many book designers insist that margins of this order are desirable for aesthetic reasons, and that because of a 'part-whole proportion illusion', an area of type matter occupying 50% of the page area will seem much larger. Research by Burt et al. (1955) on readers' preferences for margins suggested that they do indeed prefer relatively generous margins. Burt suggested that one third of the width of the page should be occupied by the margins. Tinker and Paterson, however, could find no significant difference between margins of 7/8 inch (2.2 cm) and 1/16 inch (0.16 cm) in terms of legibility. They therefore concluded that wide margins could only be justified in aesthetic terms (Tinker, 1963).

Some writers have suggested that wide margins help to prevent the eye from swinging off the page at the ends of lines and in the backward movement to the beginning of each new line. It has also been argued that generous margins help to reduce visual distraction from stimuli in the periphery of the visual field. Margins may indeed be valuable in this way where there is a marked brightness contrast between the page and any surrounding visual stimuli. Such contrasts are known to contribute to visual fatigue. In printed materials, margins also have a very practical function in providing space for the fingers to hold the book without obscuring any of the text (which is why outer and foot margins are conventionally wider than inner and head margins), and in providing space for making notes. It is also worth bearing in mind that a display area of any kind, if filled to maximum capacity, is likely to have an extremely daunting appearance. An air of spaciousness will be much more inviting for the reader.

1.10. Single- versus double-column layouts

A study by Poulton (1959) on the comprehension of scientific journals suggests that in the majority of cases a larger type size in a single column is preferable to a smaller type size in a double-column layout. With an A4 page, however, a single column of typeset text would be too wide (unless part of the page area was unused), so double-column layouts are usually adopted (Figure 5.2). In certain other situations there may be advantages in using two or three columns because a smaller type size can then be used without the lines being excessively long. For typewritten text on an A4 page a single-column layout is the most suitable. This will result in lines of a reasonable length and adequate margins (see Section 5.3.2).

The position of tables and figures must be considered in relation to the layout of the text. A study of the effect of centred tables on the legibility of single- and double-column formats in scientific journals suggests that text is more easily scanned when tabular matter is set within the width of the column; two-column layouts with tables cutting across both columns were found to be confusing, and 74% of readers preferred the single-column layout (Burnhill et al., 1976).

1.11. Paper and ink

Maximum paper reflectance undoubtedly promotes maximum reading efficiency because it provides the greatest contrast between background and image. Tinker (1963) has concluded, however, that for text printed in black ink all paper surfaces, whether white or tinted, are equally legible if they have a reflectance of at least 70%.

The effects of glossy papers on legibility have been examined in several studies. High-gloss papers often result in troublesome reflections if they are read under direct lighting, but experiments by Paterson and Tinker showed that with well diffused illumination, text printed on a high-gloss paper with 86% glare was read as fast as on paper with less than 23% glare (Tinker, 1963). Operbeck (1970) found, however, that readers preferred matt papers to medium or glossy papers.

Paper opacity is also an important factor. Spencer et al. (1977a) found that the 'show-through' of print on subsequent pages which occurs with papers of low opacity can reduce legibility very markedly. The effects were particularly severe when papers of low opacity were printed on both sides and when the show-through occurred between the lines of type actually being read (Figure 1.8). Such papers should therefore be printed on one side only, and remote areas. Both the forest and savanna zones were formerly rich in fauna, but these have been greatly depleted by widespread and indiscriminate hunting as well as by farming activities. Hunting as an occupation has consequently seriously declined. The principal animals hunted include antelopes, buffaloes, wild hogs, and certain kinds of monkeys. An activity which is closely related to hunting and trapping is the collection of snails. These are widely consumed in the forest zone and form an important item in the local diet.

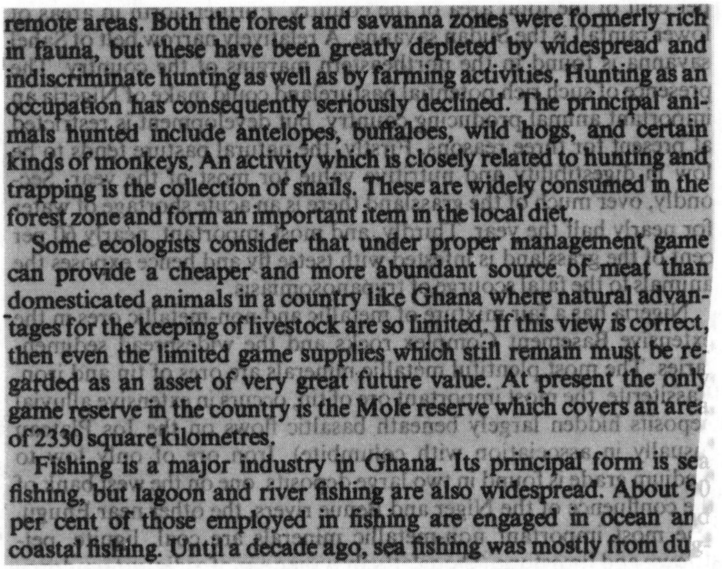

Some ecologists consider that under proper management game can provide a cheaper and more abundant source of meat than domesticated animals in a country like Ghana where natural advantages for the keeping of livestock are so limited. If this view is correct, then even the limited game supplies which still remain must be regarded as an asset of very great future value. At present the only game reserve in the country is the Mole reserve which covers an area of 2330 square kilometres.

Fishing is a major industry in Ghana. Its principal form is sea fishing, but lagoon and river fishing are also widespread. About 90 per cent of those employed in fishing are engaged in ocean and coastal fishing. Until a decade ago, sea fishing was mostly from du

Figure 1.8 Show-through.

the lines of type on successive pages should be aligned.

Ink density will also affect legibility. Experiments by Poulton (1969) with inks of different densities suggested that ink density should be at least 0.4, giving a contrast ratio of at least 60%. Operbeck (1970) found that legibility was reduced by glossy ink, and particularly when it was printed on glossy paper.

2. The presentation of text

2.1. The visual representation of information 'structure'

In the previous chapter, we were concerned with the 'legibility' of letters and words, either singly or in the form of continuous text. By legibility was meant the ease and speed with which the letters and words could be recognised and the text read. In this chapter and the two which follow it, we are concerned with the ease and speed with which a display of information can be understood. The individual elements of a piece of text, a table or an illustration may all be legible, but viewed together they may be difficult to read, to scan and to relate to one another.

In order to be easily understood, a display of information must have a logical structure which is appropriate for the user's knowledge and needs, and this structure must be clearly represented visually. In order to indicate structure, it is necessary to be able to emphasise, divide and relate items of information. Visual emphasis can be used to indicate a hierarchical relationship between items of information, as in the case of systems of headings and subheadings for example. Visual separation of items can be used to indicate that they are different in kind or are unrelated functionally, and similarly a visual relationship between items will imply that they are of a similar kind or bear some functional relation to one another. This kind of visual 'coding' helps the reader to appreciate the extent and nature of the relationship between items of information, and to adopt an appropriate scanning strategy.

These visual distinctions can be made by the systematic use of space or 'spatial coding', and by the use of typographic variations or 'typographic coding'. Sometimes it is appropriate to use one method of coding alone, and sometimes both at once.

The relative importance of items of information is often indicated by means of typographic variations such as changes in type size or weight, changes between capitals and lower case, or changes between roman and italics. In the case of headings, these typographic distinctions are often supported by a system of spatial coding based on line spacing and indentation. Distinctions between different kinds of information may be made typographically, for example by the use of italics to distinguish a quotation or a title, or they may be made spatially as is the case when information is separated into columns in a table. Functional relationships between items of information are usually made by means of spatial coding. Items placed close together will be interpreted as having a closer relationship than those which are farther apart. Rules are often used to complement spatial coding as a means of dividing and relating information, especially in tables (see Chapter 3).

Spatial and typographic coding can thus be used to great advantage in making the structure of a piece of information immediately clear to the reader. Equally, if they are misused, they can produce a very confusing result. These general principles apply not only to text, but also to tables and illustrations. These are discussed in more detail in Chapters 3 and 4 respectively.

2.2. Continuous text

2.2.1. Headings

In most scientific texts, a hierarchical system of headings will be needed. It is important to establish at the outset how many levels are required, and to use a consistent numbering system such as that shown in Figure 2.1. If more than three levels appear to be needed, it may be necessary to restructure or rewrite portions of the text in order to avoid confusing the reader. The chosen number of levels must then be represented in a consistent fashion so that the relationship between successive sections of text is clear.

Headings at different levels can be 'coded' by variations in type form and by the use of space, as has already been suggested. In conventionally typeset or IBM typeset materials, changes in type size and type weight are often used. On the standard typewriter, however, there is a very limited range of typographic variation available. Underlining is a useful coding device, but it should be used with restraint because the underline is very close to the type. This can reduce legibility, particularly if the typescript has to undergo reproduction processes which are likely to thicken-up the lettering and cause it to fuse with the underlining. (This applies especially to work for negative or diazo slides.) Dotted or double underlining should never be used, as these are untidy and aesthetically displeasing. Capitals are sometimes preferable to underlining, but these should also be used sparingly for the reasons given in Sections 1.3.5. and 5.3.2. This means that in typewritten materials it is especially important that headings should be clearly distinguished from the text and from one another by means of a logical system of spacing.

8. Discussion

8. 1.   Image quality

8. 1. 1.   Text versus numerals

Comparison of Figures 13a and 13b suggests that for some typefaces
the legibility of the numerals remained unimpaired over a slightly
wider range of image degradation than the texts, and that the fall-
off in legibility was somewhat sharper for numerals than for texts.

It must be remembered, however, that numerals have a greater visual
size than lowercase letters in the same typeface, and this may have
affected the shape of the curves. Thus it seems that if numerals are
set in the same point size as related text, it is likely to be the
text which will determine the maximum acceptable levels of degradation.
But in many instances the numerals are set in a smaller size or
photo-reduced, and it is then essential that degradation should not be
sufficient to impair their legibility because of the sharp fall-off
which occurs. This fall-off is almost certainly attributable to the
lack of redundancy in numerical material; a character is either
legible or it is not, there are no contextual cues to help in
identification.

8. 1. 2.   Applicability of image quality performance curves

It must be stressed that the results given here are applicable to
9 point type only. It is, however, likely that within the range
7-12 point, linear reductions or increases in type size (such that
the ratio between point size and set width remains unchanged) would
give similar performance curves.

8. 2.   Typeface

Comparison of Figures 13a and 13b shows that the relative legibility
of the four typefaces was very similar for text and numerals. This
suggests that characteristics of the type designs common to letters
and numerals were affecting legibility. Table 2 summarises some of
the characteristics of the typefaces, and enlarged letters and
numerals in each typeface are shown in Figure 16. The relative
legibility of the typefaces at different levels of image quality is
almost certainly the result of certain of these design characteristics,
as illustrated in Figure 17.

The relative illegibility of Baskerville when thinned down is likely
to be caused by the fact that it has a smaller x-height and thinner
strokes than the other three typefaces. Rockwell, on the other hand,
has a greater x-height and thicker, almost monoline strokes. The
presence of serifs may also enhance the legibility of thinned-down
images. Univers is possibly less legible than Rockwell because of
its lack of serifs, and Times because of its variation in stroke
thickness. Reference to Figure 3 shows marked differences in the
density measurements for Baskerville and Rockwell when thinned-down.

The relative illegibility of Rockwell when thickened-up is likely to

Figure 2.1   An example of a system of numbered headings for typewritten text.

22

Each level of heading should always be preceded by and followed by a consistent amount of space, and main headings should be preceded by more space than subheadings. Ideally, no heading should be equidistant between two paragraphs; it should always be closer to the beginning of the following paragraph to which it refers than it is to the end of the preceding paragraph. In typewritten materials, however, this may not always be a practical possibility. An example of a logical system of spacing for headings in typewritten text is shown in Figure 2.1.

Where typographic coding possibilities are limited, it may be helpful to position the main headings wholly or partly in the left hand margin. If the numbers associated with the headings are always placed in the margin, this will help to distinguish between different levels of subheading, as in Figure 2.1. Complex systems of indentation of headings and associated text should be avoided however. They can be visually confusing, and they will tend to decrease the line length unacceptably. Identation of sections of text is best reserved for special situations such as lists of points within a section or paragraph, or for quotations.

As a general rule, headings should not be centred. The eyes tend to move automatically to the left hand margin at the end of each line, and centred headings are therefore likely to interrupt the smooth flow of reading. They may even be missed altogether.

2.2.2. Paragraphs

The division of text into paragraphs containing related information undoubtedly improves readability (Tinker, 1963). It has also been shown that when the same text is divided into a small number of long paragraphs or a larger number of shorter paragraphs, readers judge the text with shorter paragraphs to be easier to read (Smith and McCombs, 1971). This is attributable to the larger amount of white space on the page which results from more frequent paragraph breaks.

Paragraphs were originally indicated by paragraph marks within the text but, when the practice of beginning each paragraph on a new line was adopted, the mark was replaced by the convention of indenting the first line by a few characters. An alternative to indentation which is now increasingly used is the insertion of extra space between paragraphs. This results in a page with a much more spacious and less daunting appearance, and it is much easier to see at a glance where the subject of the text changes. Indentation of the first line of each paragraph is redundant when space is used, except in situations where the last line on a page happens to be filled and to end with a full stop. It is then not clear whether the first line on the next page represents the beginning of a new paragraph or not. This could be used as an argument for retaining indentation in addition to space between paragraphs. Alternatively, it is usually possible to clarify the situation by taking the last word of the sentence over to the next page, or by editing the wording very slightly so that this happens automatically.

2.3. Bibliographies and references

The efficiency with which bibliographies are used can be demonstrably improved by the judicious use of spatial and typographic coding. This has been shown in two experiments by Spencer et al. (1975). Both of these experiments showed that if the references in a bibliography are listed alphabetically by author, then the location of specific author entries will be easiest if the entries are clearly distinguished from one another and if the author's name is clearly distinguished from the rest of the entry. This can be achieved simply by indenting all but the first line of each entry by two or three characters, as in System 7 in Figure 2.2. The authors' names then stand out clearly and the first two or three letters can be scanned during alphabetical searching without any visual interference from the rest of the entry. The addition of a line space between entries, as in System 9 in Figure 2.2, did not improve performance in Spencer et al.'s experiment, though it might be argued that this layout is more pleasing to the eye. In long bibliographies, however, an over-generous use of space can be counter-productive since it increases the total number of pages to be searched. It will also increase paper costs.

By convention, a certain amount of typographic coding is often used in typeset bibliographies. Authors' names may be given in capitals or small capitals, book titles and journal titles are often in italics, and journal volume numbers may be in bold. The results of Spencer et al.'s second experiment, which was concerned with typeset as opposed to typewritten bibliographies, suggested that typographic coding of this kind is unlikely to help or hinder searching if the information is well coded spatially. If it is important to save space, however, typographic coding can be helpful. It must, however, be used in such a way as to complement the reader's likely search strategy. Spencer et al. asked subjects to search for authors' names and for titles in bibliographies in the eighteen styles shown in Figure 2.3. The most effective spatial coding systems for both tasks were 4 and 5, and they were not significantly improved by typographic coding. With less adequate spatial coding, however, it was an advantage to have the authors' names in capitals

24

System 1	PAGE, MICHAEL FITZGERALD. FORTUNES OF WAR. HALE. £1.90. 823.91F (B72-10444) ISBN 0 7091 2803 7 PALLAS, NORVIN. CODE GAMES. STERLING; DISTRIBUTED BY WARD LOCK. £1.05. 001.5436 (B72-09950) ISBN 0 7061 2328 x
System 2	Allman, Michael. Geological laboratory techniques. Blanford Press. £8.50. 550.28 (B72-17338) ISBN 0 7137 0559 0 Allsop, Kenneth. Adventure lit their star. Revised ed. Penguin. £0.35. 823.91F (B72-17562) ISBN 0 14 003446 3
System 3	HAIGH, Basil. Organic chemistry of nucleic acids. Part A. Plenum Press. £9.00. 547.596 (B72-10819) ISBN 0 306 37531 1 HAINING, Peter. The Channel Islands. Revised ed. New English Library. £1.50. 914.2340485 (B72-12211)
System 4	**Barrett**, Edward Joseph. Essentials of organic chemistry. Holt, Rinehart and Winston. £5.00. 547 (B72-17335) ISBN 0 03 080348 9 **Barrow,** Charles Clement. A short history of the S.
System 5	Bartlett, Kathleen. Lovers in Autumn. Hale. £1.30. 823.91F (B72-10379) ISBN 0 7091 2329 9 Bassett, Michael Gwyn. Catalogue of type, figured & cited fossils in the National Museum of Wales. National Museum
System 6	-Cartland, Barbara. A ghost in Monte Carlo. Arrow Books Ltd. £0.25. 823.91F (B72-12081) ISBN 0 09 906180 5 -Cartwright, Frederick Fox. Disease and history. Hart-Davis. £2.50. 904.5 (B72-11136) ISBN 0 246 10537 2
System 7	Edson, John Thomas. Wagons to Backsight. Hale. £1.10. 823.91F (B72-10401) ISBN 0 7091 2394 9 Efemey, Raymond. The story of the parish church of St Thomas, Dudley. 5th ed. British Publishing. Unpriced.
System 8	Cadell, Elizabeth. Bridal array. White Lion Publishers Ltd. £1.80. 823.91F (B72-17578) ISBN 0 85617 622 2 Cafferty, Bernard. Spassky's 100 best games. Batsford. £2.50. 794.159 (B72-16145) ISBN 0 7134 0362 4
System 9	Manessier, Alfred. Manessier. Adams and Dart. £10.50. 759.4 (B72-10983) ISBN 0 239 00098 6 Mangalam, J J. Mountain families in transition: a case study of Appalachian migration. Pennsylvania State
System 10	FARNHAM, Ann. Action mathematics. 5. Cassell. £0.65. (non-net) 372.73045 (B72-15925) ISBN 0 304 93803 3 FARQUHAR, Ronald M. The earth's age and geochronology. Pergamon. £2.50. 551.701 (B72-17343) ISBN 0 08 016387 4

Figure 2.2 The ten different styles for typewritten bibliographies tested by Spencer et al.

```
A. (1) Roman caps & lowercase
   (2) Roman caps & lowercase
   (3) Roman caps & lowercase
   (4) Roman numerals
```

1. Copy runs on. 3 space units between each entry.

ISBN 0 85292 069 5. Bajin, Boris. Olympic gymnastics for men and women. Prentice-Hall. £5.50. 796.41 ISBN 0 13 633925 5. Baker, C D. Lepard's metric reckoner: for cost per thousand sheets given price per kilogramme and weight in kilogrammes. Pitman. £4.00. 338.4367620942 ISBN 0 273 25242 9. Baldwin, Brenda. Skid prevention and control. R.W. Noon. £0.30. 629.283 ISBN 0 9502394 0 2. Ball, Alan. Alan Ball's international soccer annual No. 4. Pelham. £1.00.

2. 1st element of each entry begins a new line. 2nd, 3rd and 4th elements run on.

Pettman, Dorothy. Oral embryology and microscopic anatomy: a textbook for students in dental hygiene. 5th ed. Lea and Febiger; Kimpton. £3.80. 611.314 ISBN 0 8121 0376 9.
Philips, Francis Edward. Greek philosophical terms: a historical lexicon. New York University Press; University of London Press. £3.80. 180 ISBN 0 340 09412 5.
Piatek, Clare Gray. Perspectives in surgery. Lea and Febiger; Kimpton. £9.45. 617 ISBN 0 8121 0279 7.

3. All elements begin a new line.

Learmonth, Peter.
The houses we build.
Central Committee for the Architectural Advisory Panels.
£0.25.
721.0942 ISBN 0 9502302 0 0.
Leary, Harold W.
The PL/1 machine: an introduction to programming.
Addison-Wesley. £5.60.
001.6424 ISBN 0 201 05275 x.

4. All elements begin a new line. 1st element of each entry full out. 2nd, 3rd and 4th elements successively indented.

Emmet, Louis Emanuel.
   Emmet's notes on perusing titles and practical conveyancing. 2nd (cumulative) supplement. 15th ed. Oyez. £3.00.
      346.420438 ISBN 0 85120 124 5.
Emsden, Leo.
   Sound of the sea.
      White Lion Publishers Ltd. £1.80.
         823.91F ISBN 0 85617 894 2.

5. Line space between entries. 2nd, 3rd and 4th elements run on.

Fielding, A J. Internal migration in England and Wales: a presentation and interpretation of 'city-region' data. Centre for Environmental Studies. £0.50. 301.3260942 ISBN 0 901350 52 4.

Figes, Eva. Konek landing. Panther. £0.35. 823.91F ISBN 0 586 03638 5.

Fincher, Norah M. Mingling, and other poems. Stockwell.

6. Line space between entries. Each element begins a new line.

Serraillier, Ian.
The clashing rocks: the story of Jason and the Argonauts.
Carousel Books. £ 0.20.
823.91J ISBN 0 552 52022 5.

Seuffert, Muir.
Devil at the door.
Hale. £1.40.
823.91F ISBN 0 7091 2907 6.

Figure 2.3 The 18 different styles for typeset bibliographies tested by Spencer et al.

B. (1) Roman caps
   (2) Italic caps & lowercase
   (3) Roman caps & lowercase
   (4) Italic numerals

C. (1) Italic caps & lowercase
   (2) Bold caps & lowercase
   (3) Roman caps & lowercase
   (4) Italic numerals

Hutchinson and Co. (Publishers) Ltd. £6.00. *230.01
ISBN 0 09 108850 x.* WALTERS, PATRICK GORDON.
The Cabinet. Revised ed. Heinemann Educational. £1.00.
354.4205 ISBN 0 435 83915 2. WALTON, BRUCE.
*Essays in social biology. Vol. 1: People, their needs,
environment, ecology.* Prentice-Hall. £4.00. *301.3108
ISBN 0 13 656835 1.* WARD, WILLIAM F. *Electronics
testing and measurement.* Macmillan. £4.50. *621.381
ISBN 0 333 12544 4.* WARNATH, ARTHUR WILLIAM.

GILL, JOHN. *The tenant.* Collins. £1.50. *823.91 F
ISBN 0 00 221843 7.*
GILLARD, R D. *Essays in chemistry. Vol. 3: 1972.*
Academic Press. £1.80. *540 ISBN 0 12 124103 3.*
GILLEN, LUCY. *Dangerous stranger.* Mills and Boon.
£0.80. *823.91F ISBN 0 263 05021 1.*
GILLESPIE, IAN ERSKINE. *Gastroenterology: an
integrated course.* Churchill Livingstone. £1.50. *616.3
ISBN 0 443 00854 0.*

SALT, JOHN.
Parents - participation and persuasion in primary education.
University of Sheffield Institute of Education. £0.15.
372.1103 ISBN 0 902831 07 0.
SANDERS, ED.
*The family: the story of Charles Manson's dune buggy
attack batallion.*
Hart-Davis. £2.50.
301.4494 ISBN 0 246 10528 3.

MELLISH, E MUDGE.
Rust and rot and what you can do about them.
   Angus and Robertson. £1.25.
      620.11223 ISBN 0 207 95436 4.
MENDELSOHN, JACOB.
   *Decision and organization: a volume in honor of Jacob
   Marschak.*
      North-Holland Publishing. Unpriced.
         330.1 ISBN 0 7204 3313 4.

BAKER, HILARY. *Oakes Park, Sheffield: the historic
home of the Bagshawe family since the year 1699.* English
Life Publications. £0.15. *914.2746 ISBN 0 85101 057 1.*

BALDWIN, CECIL HENRY. *Teaching science to the
ordinary pupil. 2nd ed.* University of London Press. £3.45.
507.12 ISBN 0 340 15583 3.

BALL, GEORGE SAYERS. *Who is a white-collar*

CERVINE, JO.
X-Ray diagnosis positioning manual.
Glencoe Press; Collier-Macmillan. £1.50.
616.07572 ISBN 0 02 473270 2.

CHADWICK, ANGELA.
The infernal desire machines of Doctor Hoffman: a novel.
Hart-Davis. £1.95.
823.91F ISBN 0 246 10545 3.

Bristol and Gloucestershire. Darwen Finlayson. £3.20.
914.241 ISBN 0 85208 065 4. Ramchand, Kenneth. **The
West Indian novel and its background.** Faber and Faber
Ltd. £1.70. *823.009 ISBN 0 571 10139 9.* Ramsay, Anna
Augusta Whittal. Sir Robert Peel. Constable. £3.25.
942.0810924 ISBN 0 09 458290 4. Randall, Christine.
Creating with papier-mache. Crowell-Collier; Collier-
Macmillan. £1.05. *745.54 ISBN 0 02 767190 9.* Ranis,
Gustav. **The gap between rich and poor nations:**

Perraton, Jean. **Urban systems: collection and management
of data for a complex model.** University of Cambridge
Department of Architecture. Unpriced. *301.36094229
ISBN 0 903248 29 8.*
Perry, Eric Akers. **The Parish Church of Holy Trinity,
Wickwar. New ed.** British Publishing. Unpriced. *914.241
ISBN 0 7140 0677 7.*
Perryman, Albert Charles. **Life at Brighton locomotive
works, 1928-1936.** Oakwood Press. £0.90. *625.261*

Hindmarch, Jack.
Multiple choice questions for intermediate economics.
Macmillan. £0.50.
330.076 ISBN 0 333 13570 9.
Hislop, George.
**Let history judge: the origins and consequences of
Stalinism.**
Macmillan. £5.75.
947.08420924 ISBN 0 333 13409 5.

Medlen, Wolf.
   **The Samson riddle: an essay and a play, with the text of
   the original story of Samson.**
      Vallentine, Mitchell and Co. Ltd. £2.25.
         822.914 ISBN 0 85303 152 5.
Mehmet, George Byron.
   A.B.C.'s of transistors. 2nd ed. reprinted.
      Foulsham. £1.25.
         621.381528 ISBN 0 572 00579 2.

Adams, Arlon T. **Topics in intersystem electromagnetic
compatability.** Holt, Rinehart and Winston. £12.00.
621.38411 ISBN 0 03 085342 7.

Adeney, Carol. **This morning with God: a daily devotional
guide for your quiet time. Vol. 1.** Hodder and Stoughton.
£0.40. *242.2 ISBN 0 340 15997 9.*

Adkins, Arthur William Hope. **Moral values and political**

Chamberlain, Peter.
German army semi-tracked vehicles, 1939-45.
Model and Allied Publications. £0.40.
623.747 ISBN 0 85344 136 7.

Chance, June E.
Applications of a social learning theory of personality.
Holt, Rinehart and Winston. £6.05.
155.2 ISBN 0 03 083183 0.

(A) A typewritten bibliography with the three elements of each entry separated spatially.

1. Bassingthwaighte JB, Strandell T and Holloway GA Jr (1968)
      Estimation of coronary blood flow by washout of diffusible
      indicators
      Circulation Research 1968 23 259-278
2. Edelman LS and Leibman J (1959)
      Anatomy of body water and electrolytes
      American Journal of Medicine 1959   ?56-277
3. Gaudino M and Levitt MF (1949)
      Inulin space as a measure of extra-cellular fluid
      American Journal of Physiology 1949 157 387-393
4. Jacquez JA (1972)
      Compartmental analysis in biology and medicine: Kinetics of
      distribution of tracer-labelled materials
      Amsterdam: Elsevier Publishing Co 1972 1-237
5. Zierler KL (1958)
      A simplified explanation of the theory of indicator-dilution
      for measure. ?nts of fluid flow and volume and other
      distributive phenomena
      Johns Hopkins.Hospital Bulletin 1958 103 199-217
6. Zierler KL (1965)
      Equations for measuring blood flow by external monitoring of
      radioisotopes
      Circulation Research 1965 16 309-321

(B) The same layout with a line space between entries.

1. Bassingthwaighte JB, Strandell T and Holloway GA Jr (1968)
      Estimation of coronary blood flow by washout of diffusible
      indicators
      Circulation Research 1968 23 259-278

2. Edelman LS and Leibman J (1959)
      Anatomy of body water and electrolytes
      American Journal of Medicine 1959 27 256-277

3. Gaudino M and Levitt MF (1949)
      Inulin space as a measure of extra-cellular fluid
      American Journal of Physiology 1949 157 387-393

4. Jacquez JA (1972)
      Compartmental analysis in biology and medicine: Kinetics of
      distribution of tracer-labelled materials
      Amsterdam: Elsevier Publishing Co 1972 1-237

5. Zierler KL (1958)
      A simplified explanation of the theory of indicator-dilution
      for measurements of fluid flow and volume and other
      distributive phenomena
      Johns Hopkins Hospital Bulletin 1958 103 199-217

6. Zierler KL (1965)
      Equations for measuring blood flow by external monitoring of
      radioisotopes
      Circulation Research 1965 16 309-321

Figure 2.4   Alternative presentations for bibliographies.

28

(C) A typeset bibliography illustrating the use of bold type to emphasise all titles.

1.   Bassingthwaighte JB, Strandell T and Holloway GA Jr (1968)
     Estimation of coronary blood flow by washout of diffusible indicators. Circula-
     tion Research 1968 **23** 259-278
2.   Edelman LS and Leibman J (1959)
     Anatomy of body water and electrolytes. American Journal of Medicine 1959 **27**
     256-277
3.   Gaudino M and Levitt MF (1949)
     Inulin space as a measure of extra-cellular fluid. American Journal of Physiology
     1949 **157** 387-393
4.   Jacquez JA (1972)
     **Compartmental analysis in biology and medicine: Kinetics of distribution of
     tracer-labelled materials.** Amsterdam: Elsevier Publishing Co 1972 1-237
5.   Zierler KL (1958)
     **A simplified explanation of the theory of indicator-dilution for measurements of
     fluid flow and volume and other distributive phenomena.** Johns Hopkins Hospital
     Bulletin 1958 **103** 199-217
6.   Zierler KL (1965)
     Equations for measuring blood flow by external monitoring of radioisotopes.
     Circulation Research 1965 **16** 309-321

(D) The same presentation with a space between entries.

1.   Bassingthwaighte JB, Standell T and Holloway GA Jr (1968)
     Estimation of coronary blood flow by washout of diffusible indicators. Circula-
     tion Research 1968 **23** 259-278

2.   Edelman LS and Leibman J (1959)
     Anatomy of body water and electrolytes. American Journal of Medicine 1959 **27**
     256-277

3.   Gaudino M and Levitt MF (1949)
     Inulin space as a measure of extra-cellular fluid. American Journal of Physiology
     1949 **157** 387-393

4.   Jacquez JA (1972)
     **Compartmental analysis in biology and medicine: Kinetics of distribution of
     tracer-labelled materials.** Amsterdam: Elsevier Publishing Co 1972 1-237

5.   Zierler KL (1958)
     **A simplified explanation of the theory of indicator-dilution for measurements of
     fluid flow and volume and other distributive phenomena.** Johns Hopkins Hospital
     Bulletin 1958 **103** 199-217

6.   Zierler KL (1965)
     Equations for measuring blood flow by external monitoring of radioisotopes.
     Circulation Research 1965 **16** 309-321

when searching for authors, and to have the titles in bold when searching for titles.

In typewritten bibliographies the possibilities for typographic coding are extremely limited however. Distinctions between the author, the title and the rest of the entry can then best be made by beginning each of these elements on a new line. It is likely to be an advantage to begin the title on a new line in any case, because the constant position will be helpful in scanning.

The sequence of the various elements in a reference varies according to the system adopted. For details of accepted systems in current use, British Standard 1629 (1976) should be consulted. It is often helpful to give the date of publication immediately after the author's name when the name and date are quoted in the text. There is, however, an argument for repeating the year in the journal citation, since in looking for a journal in a library it is usual to search first for the appropriate year, and then to check the volume and issue number. The convention of setting book titles and journal titles in italics, although logical, is of questionable value, since the reader is more likely to be scanning through a list of references looking at the titles of the journal articles rather than at the titles of the journals themselves. Abbreviation of journal titles should be avoided if possible, particularly when writing for an audience who may not be familiar with the journal in question. It is also preferable to choose a method of citation for journal references which avoids fussy punctuation.

Sample presentations for typewritten and typeset bibliographies which take into account the points made above are shown in Figure 2.4.

2.4. Indexes

Very little research has been carried out on the layout of indexes. Burnhill et al. (1977) have carried out one of the few investigations, and they tested three styles of index shown in Figure 2.5. Style 1 was a traditional 'balanced' approach with page numbers ranged right, Style 2 was a left-ranging version with page numbers immediately following subheadings, and Style 3 was a left-ranging version with sub-items running on. They found no significant differences between these styles in terms of ease of use. One would think, however, that it would be easier to relate page numbers to subheadings in Style 2 as compared to Style 1, and easier to scan subheadings in Styles 1 and 2. Styles 2 and 3 are quicker to produce than Style 1, and Style 3 uses less space. Style 2 or 3 would therefore seem to be preferable to Style 1.

In contents lists there is a strong argument for positioning page numbers immediately after headings rather than on the far right-hand side of the page.

Alternatively they may be placed to the left of the headings. Both of these arrangements will be easier to produce than the traditional style with right-ranged numbers, and they will also be easier to use.

Style 1		Style	Style 3
Flour beetle: see Tribolium		Flour beetle: see Tribolium	Flour beetle: see Tribolium
Fluoridation		Fluoridation	Fluoridation, investigations 142
investigations	142-3	investigations 142-3	objections 143
objections	143	objections 143	Fluoride ions 142-3
Fluoride ions	142-3	Fluoride ions 142-3	Fluorine 131
Fluorine	131	Fluorine 131	Fly agaric toadstool 32
Fly agaric toadstool	32	Fly agaric toadstool 32	Food 21, 28; appearance 85, 86;
Food	21, 28	Food 21, 28	cartoon 83; colouring, artificial
appearance	85, 86	appearance 85, 86	86; comparison 85, 86; cost 85,
cartoon	83	cartoon 83	86; fresh, compared with
colouring, artificial	86	colouring, artificial 86	preserved 85, 86; preparation
comparison	85, 86	comparison 85, 86	85, 86; preservatives in 86;
cost	85, 86	cost 85, 86	preserved 82-6; production
fresh, compared with		fresh, compared with preserved	141-2; storage 85; taste 85, 86;
preserved	85, 86	85, 86	texture 85; transport 85;
preparation	85, 86	preparation 85, 86	U.K. consumers expenditure
preservatives in	86	preservatives in 86	chart 84
preserved	82-6	preserved 82-6	Force 9, 10, 11
production	141-2	production 141-2	Forces, electrostatic: see
storage	85	storage 85	electrostatic forces
taste	85, 86	taste 85, 86	Forest: coniferous 31; deciduous
texture	85	texture 85	32; evergreen, map 31; photo.
transport	85	transport 85	tropical, map 31
U.K. consumers expenditure		U.K. consumers expenditure	Foucalt 106-7
chart	84	chart 84	Fractional distillation 89
Force	9, 10, 11	Force 9, 10, 11	Franklin, Benjamin 123, 125
Forces, electrostatic: see		Forces, electrostatic: see	Fresh, Hermann 108-110
electrostatic forces		electrostatic forces	Fritschy, Michael 76

Figure 2.5   The three styles of index tested by Burnhill et al. (after Burnhill et al. 1977).

3. The presentation of tables

3.1. Structure

The ease and speed with which tables can be understood depends very much on the tabulation logic. The author must ask himself what information the reader already has when he consults a particular table, and what information he is seeking from it. The row and column headings should relate to the information he already has, thus leading him to the information he seeks which is displayed in the body of the table (Wright, 1977). The importance of using appropriate tabulation logic is illustrated in Figures 3.1A and 3.1B.

A

		Consumer Acceptability	
		Good	Poor
High transportation costs	Fast processing time	B	[E]
	Slow processing time	A	
Low transportation costs	Fast processing time	[C]	F
	Slow processing time		G

B

Raw material	Transportation costs	Processing time	Consumer acceptability
A	High	Fast	Good
B	High	Slow	Good
C	Low	(?)	Good
E	High	(?)	Poor
F	Low	Fast	Poor
G	Low	Slow	Poor

Figure 3.1A,B These two tables show alternative ways of presenting the same information. Table *A* will be more useful for readers wishing to know which raw materials have certain characteristics; Table *B* is more suitable when they wish to find out the characteristics of particular raw materials (after Wright 1977).

The structure of the table should be as simple as possible, particularly if it is to be displayed in slide form. Many members of the general public apparently expierence difficulty in understanding two-dimensional tables, or two-way matrices as they are sometimes called (Wright, 1977). A one-dimensional presentation is therefore likely to be more effective for non-specialist audiences. Wright and Fox (1970) carried out a series of tests on tables for converting pre-decimalisation British currency in pounds, shillings and pence (£ sd) to decimal currency in pounds and new pence (£ p). They found that conversions were quicker and errors fewer with a simple table which was essentially one long column (Figure 3.2) than with a more complex arrangement where the rows gave the values for multiples of a shilling and the columns gave the values for 0 to 11 pence (Figure 3.3).

Frase (1969) has stressed the importance of presenting the information in a sequence which will be compatible with the reader's strategy of processing. This usually operates from left to right and from top to bottom.

OLD s. d.	NEW PENCE	OLD s. d.	NEW PENCE	OLD s. d.	NEW PENCE	OLD s. d.	NEW PENCE	OLD s. d.	NEW PENCE
		2/-	10p	4/-	20p	6/-	30p	8/-	40p
1d	½p	2/1d	10½p	4/1d	20½p	6/1d	30½p	8/1d	40½p
2d	1p	2/2d	11p	4/2d	21p	6/2d	31p	8/2d	41p
3d	1p	2/3d	11p	4/3d	21p	6/3d	31p	8/3d	41p
4d	1½p	2/4d	11½p	4/4d	21½p	6/4d	31½p	8/4d	41½p
5d	2p	2/5d	12p	4/5d	22p	6/5d	32p	8/5d	42p
6d	2½p	2/6d	12½p	4/6d	22½p	6/6d	32½p	8/6d	42½p
7d	3p	2/7d	13p	4/7d	23p	6/7d	33p	8/7d	43p
8d	3½p	2/8d	13½p	4/8d	23½p	6/8d	33½p	8/8d	43½p
9d	4p	2/9d	14p	4/9d	24p	6/9d	34p	8/9d	44p
10d	4p	2/10d	14p	4/10d	24p	6/10d	34p	8/10d	44p
11d	4½p	2/11d	14½p	4/11d	24½p	6/11d	34½p	8/11d	44½p
1/-	5p	3/-	15p	5/-	25p	7/-	35p	9/-	45p
1/1d	5½p	3/1d	15½p	5/1d	25½p	7/1d	35½p	9/1d	45½p
1/2d	6p	3/2d	16p	5/2d	26p	7/2d	36p	9/2d	46p
1/3d	6p	3/3d	16p	5/3d	26p	7/3d	36p	9/3d	46p
1/4d	6½p	3/4d	16½p	5/4d	26½p	7/4d	36½p	9/4d	46½p
1/5d	7p	3/5d	17p	5/5d	27p	7/5d	37p	9/5d	47p
1/6d	7½p	3/6d	17½p	5/6d	27½p	7/6d	37½p	9/6d	47½p
1/7d	8p	3/7d	18p	5/7d	28p	7/7d	38p	9/7d	48p
1/8d	8½p	3/8d	18½p	5/8d	28½p	7/8d	38½p	9/8d	48½p
1/9d	9p	3/9d	19p	5/9d	29p	7/9d	39p	9/9d	49p
1/10d	9p	3/10d	19p	5/10d	29p	7/10d	39p	9/10d	49p
1/11d	9½p	3/11d	19½p	5/11d	29½p	7/11d	39½p	9/11d	49½p

Figure 3.2   An explicit conversion table which is essentially one long column (after Wright and Fox, 1970).

		1	2	3	4	5	6	7	8	9	10	11
Old s. d.		1	2	3	4	5	6	7	8	9	10	11
NEW PENCE		½	1	1	1½	2	2½	3	3½	4	4	4½
OLD s. d.	2/-	2/1	2/2	2/3	2/4	2/5	2/6	2/7	2/8	2/9	2/10	2/11
NEW PENCE	5	5½	6	6	6½	7	7½	8	8½	9	9	9½
OLD s. d.	2/-	2/1	2/2	2/3	2/4	2/5	2/6	2/7	2/8	2/9	2/10	2/11
NEW PENCE	10	10½	11	11	11½	12	12½	13	13½	14	14	14½
OLD s. d.	3/-	3/1	3/2	3/3	3/4	3/5	3/6	3/7	3/8	3/9	3/10	3/11
NEW PENCE	15	15½	16	16	16½	17	17½	18	18½	19	19	19½
OLD s. d.	4/-	4/1	4/2	4/3	4/4	4/5	4/6	4/7	4/8	4/9	4/10	4/11
NEW PENCE	20	20½	21	21	21½	22	22½	23	23½	24	24	24½
OLD s. d.	5/-	5/1	5/2	5/3	5/4	5/5	5/6	5/7	5/8	5/9	5/10	5/11
NEW PENCE	25	25½	26	26	26½	27	27½	28	28½	29	29	29½
OLD s. d.	6/-	6/1	6/2	6/3	6/4	6/5	6/6	6/7	6/8	6/9	6/10	6/11
NEW PENCE	30	30½	31	31	31½	32	32½	33	33½	34	34	34½
OLD s. d.	7/-	7/1	7/2	7/3	7/4	7/5	7/6	7/7	7/8	7/9	7/10	7/11
NEW PENCE	35	35½	36	36	36½	37	37½	38	38½	39	39	39½
OLD s. d.	8/-	8/1	8/2	8/3	8/4	8/5	8/6	8/7	8/8	8/9	8/10	8/11
NEW PENCE	40	40½	41	41	41½	42	42½	43	43½	44	44	44½
OLD s. d.	9/-	9/1	9/2	9/3	9/4	9/5	9/6	9/7	9/8	9/9	9/10	9/11
NEW PENCE	45	45½	46	46	46½	47	47½	48	48½	49	49	49½

Figure 3.3 A table with the information arranged horizontally rather than vertically (after Wright and Fox, 1970).

Wright and Fox (1970) found that conversions from £p to £sd were faster if £p values were on the left, and vice versa. Earlier studies had shown, however, that provided subjects were asked to make conversions in one direction only, then it made no difference whether the known information was in the left-hand column or the right-hand column.

Ideally, any items of information within a table which need to be compared with one another should be listed vertically. Tinker (1965) has shown that searching for a single item in a randomly ordered list is quicker if the items are listed vertically one beneath the other, rather than horizontally. Similarly, Woodward (1972) has shown that when subjects are asked to compare pairs of numbers in a table and identify them as the same or different, they are able to make the comparison more quickly if the numbers are printed one above the other than if they are printed end to end. Thus, in the kind of table shown in Figure 3.4 which gives values for six different attributes for each of six different items, the decision as to whether to assign the items to the rows and the attributes to the columns, or vice versa, should ideally be based on the way in which the table will be used. If the reader is most likely to want to look up a particular item and then read off the values for its attributes, the items should be listed vertically so as to form the rows. If, on the other hand, the reader is more likely to want to look up an attribute and then read off the value for each item then the attributes should be listed vertically to form the rows.

34

(A)

Items	Attributes					
	1	2	3	4	5	6
A	28.05	00.31	02.93	06.04	15.39	07.80
B	31.59	00.07	03.92	09.28	16.48	08.39
C	26.83	00.24	02.63	07.22	14.75	07.32
D	27.62	00.12	03.01	07.89	15.42	06.94
E	26.45	00.36	02.75	08.62	14.26	07.75
F	26.83	00.09	03.03	07.52	14.79	07.67

(B)

Attributes	Items					
	A	B	C	D	E	F
1	28.05	31.59	26.83	27.62	26.45	26.83
2	00.31	00.07	00.24	00.12	00.36	00.09
3	02.93	03.92	02.63	03.01	02.75	03.03
4	06.04	09.28	07.22	07.89	08.62	07.52
5	15.39	16.48	14.75	15.42	14.26	14.79
6	07.80	08.39	07.32	06.94	07.75	07.67

Figure 3.4 Version (*A*) of this table is likely to be the most useful, since it facilitates alphabetical scanning of the list of items and comparison of attributes across items.

In many cases, however, the decision will depend on the number of items in relation to the number of attributes. Thus if there are more items than attributes, it will usually be more economical in terms of space if the items are listed vertically to form the rows.

Wherever possible, numerical tables should be explicit rather than implicit, i.e. the information should be given in full (Wright and Fox, 1970). In an implicit table, the reader may be required to add together two values in order to obtain a third which is not explicitly stated in the table. An example of such a table is shown in Figure 3.5. Implicit tables save space, but require more effort on the part of the reader and may cause confusion and errors. They are particularly unsuitable for slides and other transient displays.

3.2. Layout and typography

3.2.1. General principles

Although tables are constructed by grouping data into vertical columns, in many instances the reader will wish to select an item in the first column and then scan across the table horizontally. This will be the case where several

New Pence	Old Pence		New Pence	Old Shillings
1	2d		10	2/-
			15	3/-
2	5d		20	4/-
			25	5/-
3	7d		30	6/-
			35	7/-
4	10d		40	8/-
			45	9/-
5	1/-		50	10/-
6	1/2		55	11/-
			60	12/-
7	1/5		65	13/-
			70	14/-
8	1/7		75	15/-
			80	16/-
9	1/10		85	17/-
			90	18/-
			95	19/-
			100	£1

Figure 3.5 An implicit conversion table (after Wright and Fox, 1970).

items are listed vertically to form the rows and their various attributes form the columns, as in Figure 3.4. Although the data within each vertical column are related in kind, the information within each row has a functional relationship. It is this relationship which interests the reader and which must be emphasised in this kind of table. Horizontal and vertical scanning will be of equal importance in two-way matrices, but since proximity of rows is always easier to achieve than proximity of columns, it is again the horizontal relationships which need to be emphasised.

The basic principle which should be observed in designing tables is that of grouping related data, either by the use of space or, if necessary, rules. Items which are close together will be seen as being more closely related than items which are farther apart, and the judicious use of space is therefore vitally important. Similarly, ruled lines can be used to relate and divide information, and it is important to be sure which function is required. Rules should not be used to create closed compartments; this is time-wasting and it interferes with scanning.

3.2.2. Horizontal emphasis

Horizontal scanning can be facilitated by generous spacing between rows.

The space between columns, on the other hand, should be just sufficient to separate them clearly, but no more. The columns should not, under any circumstances, be spread out merely to fill the width of the type area. Hartley (1978) has shown that left-ranging tables are certainly no less easy to read then 'centred' tables, and they are simpler to type. Sometimes, however, it is difficult to avoid undesirably large gaps between columns, particularly where the data within any given column vary considerably in length. This problem can sometimes be solved by reversing the order of the columns, as in Figures 3.6 and 3.7. In other instances the insertion of additional space after every fifth entry or row can be helpful, as in Figures 3.8 and 3.9, but care must be taken not to imply that the grouping has any special meaning.

Where the amount of vertical space available for a table is limited, or where space alone between rows does not give sufficient horizontal emphasis, the use of thin horizontal rules can be helpful. This is illustrated in Figure 3.10.

Table xx

Distribution of places of meeting of spouses

Place of meeting	%
Dance or dance hall	27.3
Private house	17.6
Work or Forces	14.6
Street or public transport	9.7
Cafe or pub	6.1

Figure 3.6   In this table the information has been arranged about a central axis in an attempt to produce a 'balanced' appearance (after Hartley, 1978).

Table xx

Distribution of places of meeting of spouses

%	Place of meeting
27.3	Dance or dance hall
17.6	Private house
14.6	Work or Forces
9.7	Street of public transport
6.1	Cafe or pub

Figure 3.7   In this version of the table shown in Figure 3.6, the information has been arranged for maximum ease of use (after Hartley, 1978).

	Rumpsteak	Pork chops	Potatoes	Butter	Margarine	Cheese
Athens	0.70 —11	0.65 + 8	0.07 +2	0.66 + 9	0.33 — 8	0.41 + 1
Bonn	1.35 —11	0.97 + 3	0.03	0.71 + 2	0.37 + 4	1.05 +32
Brussels	1.21 + 1	0.82 +11	0.02	0.58 — 3	0.26 — 9	0.54 — 9
Copenhagen	1.47 +14	0.33 + 6	0.07 +3	0.67 +13	0.27 — 4	0.75 +14
Dublin	0.75 +20	0.82 +15	0.06 +2	0.47 +10	0.33 + 9	0.53 +13
Geneva	2.30 + 9	1.30 — 3	0.09 +2	0.70 + 2	0.45 — 7	0.98 + 4
Hague	1.07	1.76 + 1	0.06	0.57	0.16 — 1½	0.72
London	1.34 +33	0.72 + 3	0.04	0.31 + 7	0.29 + 5	0.42 + 2
Luxembourg	1.30 + 9	0.65 +11	0.02	0.58 + 3	0.32 + 2	0.80
Oslo	0.94 —68	1.12 + 7	0.07 —1	0.42 + 2	0.21 — 8	0.59 +14
Paris	1.32 +25	0.83 +15	0.04 —2	0.71 + 9	0.29 + 1	0.55 +25
Rome	1.21 +10	0.85 + 8	0.04 —1½	0.85 +12	0.15 +5½	0.76 + 3
Stockholm	1.28 + 6	0. + 2	0.08 +1	0.56	0.35 —15	0.74 + 4
Vienna	1.21 + 6	0.89 + 1	0.09 +5	0.61 + 3	.34 + 2	0.58 + 8

The plus minus figures are changes in the past six months. Prices in £ per pound.

Figure 3.8   The purpose of this table is to allow comparison of food prices in London with those in other European cities (after Hartley, 1978).

	Rumpsteak	Pork chops	Potatoes	Butter	Margarine	Cheese	
London	1.34 +33	0.72 + 3	0.04	0.31 + 7	0.29 + 5	0.42 + 2	London
Athens	0.70 —11	0.65 + 8	0.07 +2	0.66 + 9	0.33 — 8	0.41 + 1	Athens
Bonn	1.35 —11	0.97 + 3	0.03	0.71 + 2	0.37 + 4	1.05 +32	Bonn
Brussels	1.21 + 1	0.82 +11	0.02	0.58 — 3	0.26 — 9	0.54 — 9	Brussels
Copenhagen	1.47 +14	0.33 + 6	0.07 +3	0.67 +13	0.27 — 4	0.75 +14	Copenhagen
Dublin	0.75 +20	0.82 +15	0.06 +2	0.47 +10	0.33 + 9	0.53 +13	Dublin
Geneva	2.30 + 9	1.30 — 3	0.09 +2	0.70 + 2	0.45 — 7	0.98 + 4	Geneva
Hague	1.07	0.76 + 1	0.06	0.57	0.16 —1½	0.72	Hague
Luxembourg	1.30 + 9	0.65 +11	0.02	0.58 + 3	0.32 + 2	0.80	Luxembourg
Oslo	0.94 —68	1.12 + 7	0.07 —1	0.42 + 2	0,21 — 8	0.59 +14	Oslo
Paris	1.32 +25	0.83 +15	0.04 —2	0.71 + 9	0.29 + 1	0.55 +25	Paris
Rome	1.21 +10	0.85 + 3	0.04 —1½	0.85 +12	0.15 +5½	0.76 + 3	Rome
Stockholm	1.28 + 6	0.91 + 2	0.08 +1	0.56	1.35 —13	0.74 + 4	Stockholm
Vienna	1.21 + 6	- 1	0.09 +5	0.61 + 3	0.34 + 2	0.58 + 8	Vienna

The plus minus figures changes in the past six months. Prices in £ per pound.

Figure 3.9   The table shown in Figure 3.8 has been made easier to use by placing London first so that comparisons with other cities can be made more easily, by repeating the names of the cities on the right, and by leaving a line space after every fifth entry to facilitate reading across from column to column (after Hartley, 1978).

	Group one	Group two	Group three	sex
Class A	1234	14	986	♂
	4567	6	431	♀
Class B	9849	84	18	♂
	8321	100	1000	♀
Class C	4679	8	864	♂
	2131	2	9	♀

no!

This layout is harder to scan and more time consuming to set in printers type, and is therefore more costly. It also occupies more space than the other two layouts.

		GROUP		
Class		I	II	III
A	♂	1234	14	986
	♀	4567	6	431
B	♂	9849	84	18
	♀	8321	100	1000
C	♂	4679	8	864
	♀	2131	2	9

?

Interrupted lines can be useful if one column requires emphasis by being set apart but too many interruptions interfere with rapid scanning.

		GROUP		
Class		I	II	III
A	♂	1234	14	986
	♀	4567	6	431
B	♂	9849	84	18
	♀	8321	100	1000
C	♂	4679	8	864
	♀	2131	2	9

yes

This is not only the most economic layout, but being the simpliest it is the easiest to scan.

Figure 3.10   Three examples of the same data using different systems of ruled lines.

3.2.3. Vertical emphasis

The relation between the information within each column must also be made clear. It is an advantage if the data in each column can be left-justified in order to create a visual grouping, but this is not always possible. Numbers which need to be added together, for example, cannot be justified towards the left. The temptation to use vertical rules between columns should be strongly resisted however, since they will tend to draw the eye down the table rather

39

than across. In some cases the use of typographic distinctions between columns can be helpful. Wright and Fox (1970) found that the use of light and bold type to distinguish between the columns in a simple conversion table reduced the number of errors made in making conversions, though it made no difference to the speed at which conversions were made. They concluded that such distinctions might be particularly helpful in multi-column tables.

3.2.4. Headings

Row and column headings should be carefully positioned so that it is immediately clear which data they refer to. Any abbreviations of units in headings should conform with established international standards. Lengthy column headings are often responsible for unnecessarily wide spacing between columns. In some cases this problem can be solved by sloping the column headings, although this is not desirable if it can be avoided. Sloped headings are difficult to position accurately using a typewriter, and are often best typed separately and added to the artwork at a later stage by the cut and stick method described in Section 12.3.2.

3.2.5. Lettering

Lettering on tables should conform with the standards laid down in the chapters on particular media. Where tables are typeset for printing, it is not advisable that they should appear in a smaller type size than the accompanying text, particularly if the information is likely to be copied or microfilmed subsequently.

4. The presentation of graphs, charts and diagrams

4.1. The relative merits of different kinds of illustration

When should a graph be used as well as, or instead of, a table? Carter (1947) suggests that if the reader needs precise numerical information, this can be obtained more rapidly and accurately from a table. Graphs are helpful, however, where an interpretation of the relation between the variables is important.

Charts of various kinds are sometimes a useful alternative to tables or graphs. Wright (1977) reports that for the general public bar charts may be more comprehensible than tables for certain purposes. She also cites studies which suggest that bar graphs may be preferable to line graphs where the data can be subdivided in a number of different ways and the reader is required to make static comparisons. It would seem, however, that line graphs are superior where dynamic comparisons of data plotted against time are required. Schutz (1961b) found that where extrapolation of a trend was required, line graphs were better than vertical bar charts, and vertical bar charts were better than horizontal bar charts.

Where several different variables are to be presented in a bar chart, the information can usually be arranged in a number of different ways (see Figure 4.1). The ease of comparison of different variables will depend on the arrangement chosen. The author must therefore decide which comparisons the reader is likely to want, then design the chart accordingly.

Pie charts may also be useful in some instances. Wright (1977) cites a study carried out in 1929 which suggested that a pie chart may be superior to a bar chart where three or four comparisons are being made, but for accuracy a table would be superior to both. Hailstone (1973) has suggested that charts or symbols which rely on proportional area to denote quantity may not be entirely satisfactory as most people tend to make comparisons on a linear basis.

Relatively little research has been carried out on the effectiveness of diagrams other than graphs and charts. Dwyer (1976) has found that although photographs may be superior in terms of immediate retention of

41

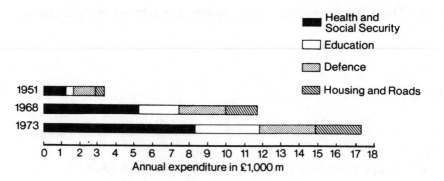

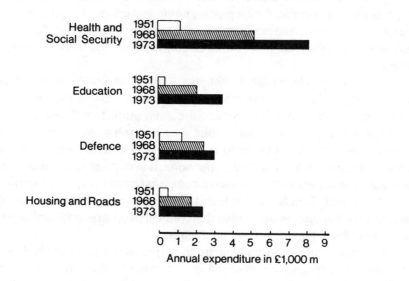

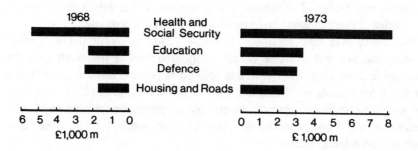

Figure 4.1   Three alternative presentations of the same information using horizontal bar charts. Different arrangements of the information facilitate different comparisons (after Wright, 1977).

information, line diagrams are better for delayed retention. He found that simple line diagrams in colour were the most effective form of illustration in a text on the functioning of the heart. The addition of realistic detail to the diagram did not automatically improve it, and excessive detail tended to reduce its effectiveness. Students preferred coloured illustrations, even when the colour conveyed no additional information.

A useful survey of various kinds of diagram which may be used for particular purposes is given by Lockwood (1969).

4.2. The content of illustrations

The content of an illustration and the way in which the information is presented must be determined in relation to the medium for which it is intended, and the kind of audience. A detailed illustration which is satisfactory for a printed publication may be entirely unsuitable for presentation as a slide. (This point is discussed in more detail in Chapter 6.) Furthermore, a style of presentation which is suitable for one publication may not be suitable for another, and a slide intended for a scientific meeting will not necessarily be suitable for a non-specialist audience. This must be borne in mind when planning an illustration or a series of illustrations.

Any illustration must have an ordered structure based on a clear set of priorities, and it will be most effective when its structure and the data it contains are simple rather than complex. Unnecessary items of information will merely detract from the illustration by increasing its overall complexity. Some scientific findings are unavoidably complex however. In studies relating to the clinical aspects of haematology, for example, many factors may need to be considered at once. In this case, it may be preferable to build up the data by means of a series of related illustrations, particularly if the information is in slide form. The complex overall picture may then be shown as a final illustration. This method will help the audience or readers to understand the data while making no attempt to pretend that the overall picture is less than complex.

In the case of graphs, the number of lines which can be included on any one illustration will depend largely on how close the lines are and how often they cross one another. Three or four is likely to be the maximum acceptable number. In some instances, there may be an argument for using several graphs with one line each as opposed to one graph with multiple lines. It has been shown that these two arrangements are equally satisfactory if the user wishes to read off the value of specific points; if, however, he wishes to compare the lines, than the single multi-line graph is superior (Schutz, 1961a).

Each and every item in an illustration should have a direct and simple purpose, and this rule includes labels and notations of various kinds. Nothing should be added if it does not contribute to the meaning of the illustration.

4.3. General principles of presentation

4.3.1. Layout

Since European languages are written from left to right and from the top of the page to the bottom, the eye naturally tends to alight first in the top left hand corner of any page (Figure 4.2), unless there are strong compositional features which draw it elsewhere. This is the most important area of any illustration therefore. As far as charts and graphs are concerned, the left and lower margins are also important. In planning this kind of illustration, it is convenient to divide the space as illustrated in Figure 4.3. A rectangle two units by three units has been sectioned so that the long dimension is divided into six parts and the short dimension into four parts. This is not a hard and fast rule which should necessarily be adhered to at all times, but it provides a useful basis for many kinds of scientific illustration and it allows enough marginal space for the labelling of axes on graphs and charts.

4.3.2. Framing

The practice of framing an illustration with a drawn rectangle is not recommended. This kind of typographic detailing should never be added purely for aesthetic reasons or for decoration. A simple, purely functional drawing will automatically be aesthetically pleasing. Unnecessary lines usually reduce both legibility and attractiveness. It must also be remembered that although the illustrator sees the artwork unframed, this is *not* the final product. All work for printed publications will ultimately be framed, either by adjacent type, adjacent white margins or the edges of the page. All work for slides will be framed by the slide mount. If framing is drawn on the artwork then a double frame will be created, thus over-emphasising an unimportant area of the illustration.

4.3.3. Visual emphasis

As a general rule, plotted points and graph lines should be given more 'weight' than the axes. In this way the 'meat' will be easily distinguishable from the 'bones'. Furthermore, an illustration composed of lines of unequal weights is always more attractive than one in which all the lines are of uniform thickness. It may not always be possible to emphasise the data in this way,

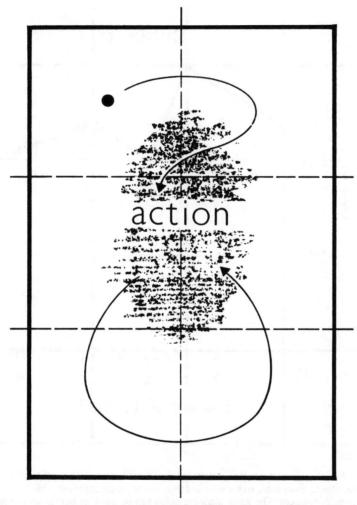

Figure 4.2 In most western cultures the observer's eye starts scanning a page from a point in the top left sector. A sweeping search of the page is then made, finishing where the main action is expected to be. The latter is normally found in the upper third of the central sector.

The design of the illustration may encourage or oppose this natural tendency. Opposing designs can be slower to interpret and sometimes confusing.

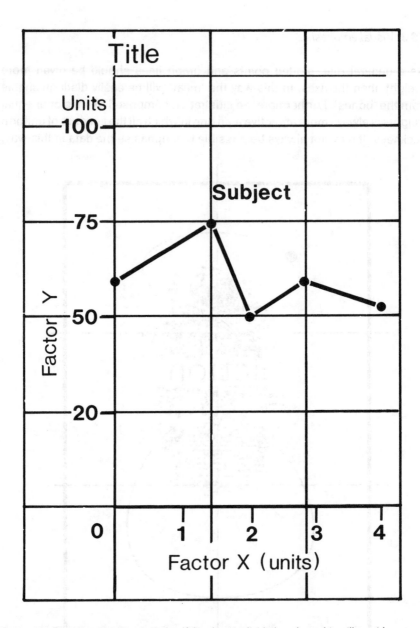

Figure 4.3 Planning consistent graphs. If the sheet is divided as above this will provide enough space at the left and lower margins for axes and their labels. Space at the right-hand margin can be as small as convenient. The same planning guide can be used for horizontal (landscape) format graphs.

however. In a scattergram, for example, the more plotted points there are, the smaller they may need to be and this will give them a lighter appearance. Similarly, the more curves there are on a graph, the thinner the lines may need to be. In both cases, the axes may look better if they are drawn with a somewhat bolder line so that they are easily distinguishable from the data.

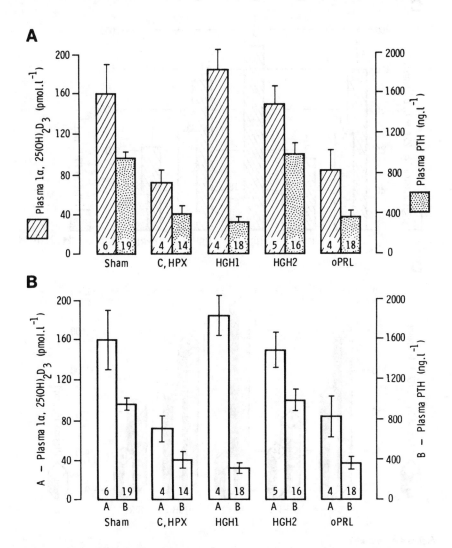

Figure 4.4 Four ways of presenting the same data.
   A. A conventional column chart. The artist must be sure that the left- and right-hand scales refer to the left- and right-hand columns respectively. Standard error bars are not drawn inside the columns in order to improve clarity.
   *B.*The effect without shading is visually less dramatic.

Cross hatching and other forms of shading should almost always be composed of thinner lines than those used for the rest of the drawing (Figure 4.4).

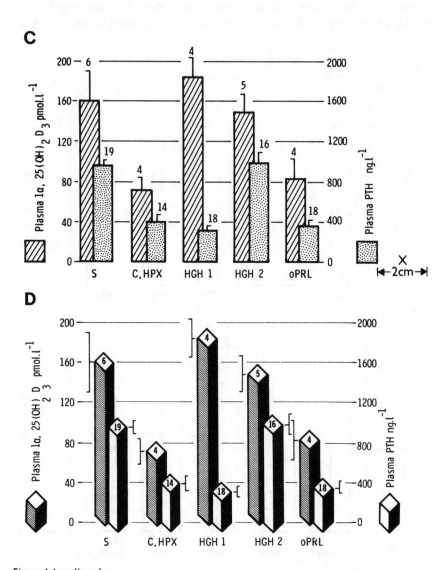

Figure 4.4 continued.

C. In this example the columns are overlapping, thus saving 2 cm of horizontal space as compared with design *A.* The standard error bar terminals are treated differently in order to avoid neighbouring columns.

D. The three dimensional design allows the population for each column to be shown at the column ends. If used for colour slides this technique can be very attractive.

48

4.4 The presentation of graphs and charts

4.4.1. Data points

The plotted points on a graph should always be made to stand out well. They are, after all, the most important feature of a graph, since any lines linking them are nearly always a matter of conjecture. These lines should stop just short of the plotted points so that the latter are emphasised by the space surrounding them. Where a point happens to fall on an axis line, the axis should be broken for a short distance on either side of the point.

If a multiple line presentation is used, it may be helpful to code the points by different geometric shapes, either as well as or instead of coding the lines themselves in some way. Variations of this kind should be kept to a minimum, however, or they will become untidy and confusing. As a general principle it is always better to opt for simplicity.

4.4.2. Standard errors

Standard errors of the mean are usually drawn as lines extending vertically above and below a plotted point or the top of a column. Normally these lines are of equal length and therefore only one of the pair needs to be drawn. The choice will depend on the circumstances. A multiple-line graph can look very cluttered if all the standard error lines are drawn in.

Standard error lines look best if drawn in a thinner line than that used for the main body of the data. Thus, if a 0.5 mm line thickness has been chosen for the graph lines, 0.35 mm will be suitable for the standard error lines. The latter should always be drawn to actually touch their associated plotted points or bars. This avoids any possibility of confusion or ambiguity.

It is usual to terminate standard error lines with a short horizontal finial. A quick and neat method of executing these is described in Section 12.2.6.

4.4.3. Lines linking data points

On many kinds of graph it is appropriate to link the data points with lines, as this will help the eye to find related points and to follow trends in the data.

In multiple-line presentations there is often some degree of overlap between the lines. This can give rise to a great deal of confusion, particularly where the data points for two or more different lines happen to fall very close together. This confusion can be greatly reduced if the lines are given an order of precedence, such that when two lines cross, the line with the lower position in the hierarchy is broken and appears to pass behind the line having precedence. The information contained in Figure 4.5 could be interpreted in several ways, but this ambiguity would not be possible if the lines were given an order of precedence.

In some cases it may be necessary to code the lines themselves in some way. This can be done by using lines of different thicknesses, or lines broken in a variety of patterns using dots and dashes. Schutz (1961b) found that colour coding of the lines was of some help on multiple-line graphs. This is not a practical proposition for artwork which is intended for publication in journals, but colour can be used in this way on graphs specifically designed for other media such as slides or overhead projector transparencies.

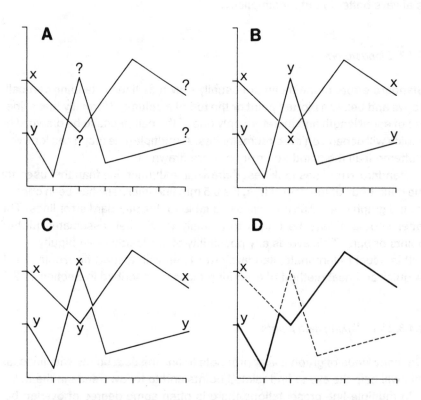

Figure 4.5 Lines without a hierarchical order give rise to confusion. Diagram *A* could be interpreted as in diagrams *B* or *C*. In fact the author did not intend either of these. Diagram *D* is not ambiguous. Not only have the graph lines been given a hierarchy, but they have not been allowed to touch each other.

4.4.4. Scale calibrations

The frequency of labelled scale calibrations on the axes of a graph can significantly affect the accuracy with which it is interpreted. As little interpolation as possible should be required of the user, in order to minimise errors. If single units cannot be marked, it has been suggested that multiples of 2, 5 or 10 should be used (Wright, 1977).

Calibrations should be clearly indicated on the axis in such a way that they will still be clearly visible when the artwork has been reduced. In later chapters it will be suggested that artwork for illustrations should ideally be drawn within a rectangle measuring 13 × 20 cm; on this basis, calibrations should be at least 3-4 mm long. These are often drawn inside the axes on the grounds that this gives a neat and aesthetically pleasing appearance to the graph, but calibrations drawn in this position may interfere with the data. For this reason we recommend that they should always be drawn on the outside of the axes (pointing towards the number to which they refer).

The treatment of scales as in Figure 4.6 can be extremely helpful in relating the position of the data to the vertical scale. This effect is achieved by removing the vertical scale altogether and extending the calibrations as thin continuous lines, thus forming a background to the data horizontally. This system works equally well for most graphs and bar charts.

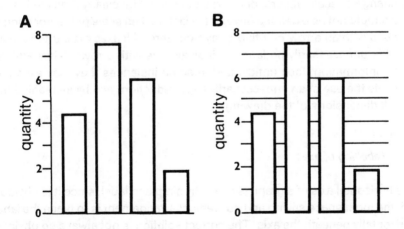

Figure 4.6 The way in which scales are treated can aid or retard comprehension. The quantity value for each bar is more quickly assessed in *B*. This diagram is also more aesthetically pleasing because the scale provides a background against which the white bars stand out more easily than in *A*.

4.4.5. Axes

The axes of a graph or chart are usually represented by a vertical line on the left and a horizontal line at the bottom of the plot. A line is not always necessary however. Space can be used to create the impression of a border or line, and this nearly always leads to aesthetically pleasing results as well as saving labour. In Figure 4.6, for example, there is no need for the baseline to be drawn in; it already exists in the eye of the observer.

The desire to superimpose several different kinds of data which have a common independent variable such as time, needs to be resisted. For example, differently measured components of the blood may behave in different ways, over a given space of time. Some scientists may ask for such data to be illustrated on superimposed axes. The result is more often than not a misleading unscientific compromise, preventing the quick interpretation of the data (Figure 4.7A,B).

The scientific press has examples of several differently measured components all superimposed (Figure 4.8A,B), with several scales both to left and right of the graph. If the research is to be taken seriously then so should the graphic means of communicating it. A simple statement of each event in turn is always more efficient and is less likely to obscure the truth.

The choice of scales for the axes should be very carefully considered. An ill-advised choice can give a very misleading impression of the data, as shown in Figure 4.9. If an axis does not start at zero (or 1 in the case of log scales), then this fact must be clearly shown by an obvious break in the scale. If a change of scale occurs, once again this must be clearly indicated. Axis lines should not be extended beyond the last marked scale point, nor should they end with an arrow pointing away from zero. All axes extend to infinity, and it is unnecessarily pedantic to indicate this with arrows. They add no useful information, look untidy, and in some instances they may even necessitate the use of an unnecessarily large reduction ratio by increasing the overall dimensions of the drawing.

4.4.6. Labelling of axes

The horizontal axis of a graph is relatively simple to label, since it is obvious that the most space saving and convenient arrangement is to place the label horizontally beneath the axis. The correct solution is not always so obvious for the vertical axis however. If the label is written vertically, this will leave the maximum area available for the plot itself, but it will mean that the label is less convenient to read. In printed materials this is not a major problem as the

52

reader can easily turn the page through 90 degrees, and he can study the graph for as long as he wishes. With slides, however, the audience will have to turn their heads sideways in order to read the label, and they will have less time to do so. Vertical labels are therefore preferable for printed materials

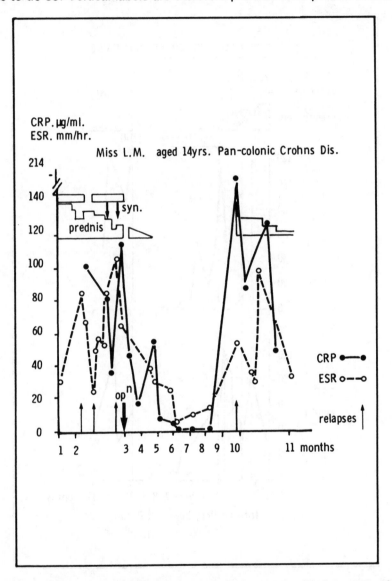

Figure 4.7A Draft layout of data as originally conceived for a 35 mm slide. It is unscientific to compare two *differently measured* variables on the *same* axis. Therefore ESR and CRP must be separated. A square format is a disadvantage since it will waste 1/3 of the slide area. Data used with permission of Dr. Fagan, RPMS London.

and horizontal ones for slides. Many illustrations are prepared for both purposes, however, and the question then arises as to which medium should take precedence. The answer should be obvious. Lectures are transient events informing only a few persons at a time, and members of the audience

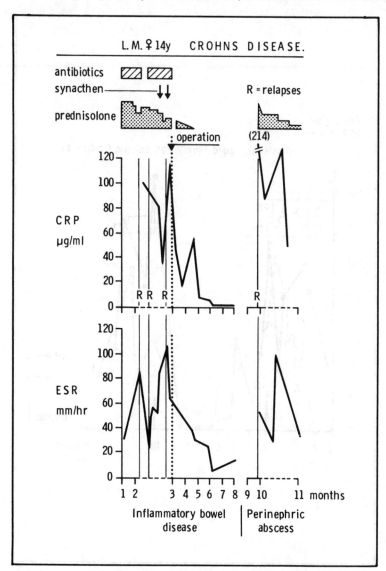

Figure 4.7B The finished drawing shows how the data were redesigned. The full 2:3 slide format has been occupied. The two differently measured y-axis variables are now clear. Common events in time coincide vertically. The two disease phases are distinct. Treatment is not obscured by other data. Six character units describe the patient instead of the former fifteen.

will usually hear a lecture once only, with no subsequent opportunity of referring to the illustrations. Printed publications, however, provide a much more permanent record which can be referred to time and time again. Published data should therefore be regarded as more important than slides, and labels which waste space on the printed page should be avoided. As a general rule, axes should be labelled only with their appropriate S.I. units of measurement and not with the subject matter of the graph (Figure 4.10).

Sometimes short vertical axis labels can be accomodated horizontally, but where this is possible, some thought must be given to their positioning. Lining up the label with the left hand margin created by the scale point numerals is often the best looking procedure, as shown in Figure 4.11. If the y (vertical) axis label is not centred, some designers prefer to put the x (horizontal) axis label on the extreme right. This may look pleasing, but it can be time consuming to execute. If stencils or dry transfer lettering are used it will be necessary to work from right to left, and a degree of skill in placing

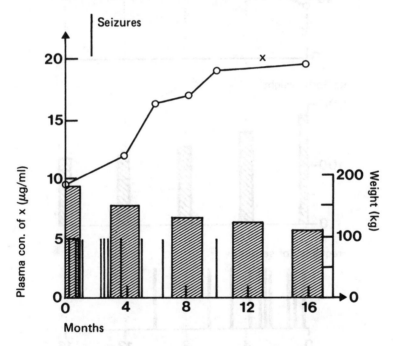

Figure 4.8A A badly designed graph, based on examples from the scientific press.

Faults: Inward facing scale calibrations add confusion to an already muddled picture. The label on the right-hand scale reads downwards. This is *always* wrong. It is not clear from the x-axis label what is being measured in months. The key for 'seizures' is disassociated from the area where seizures occur. 'Con.' on the y-axis is not a standard abbreviation. Arrows at the ends of axes are pedantic and irrelevant. The double zero at the intersection of the two axes is not necessary.

the lettering will be necessary for a pleasant result. If a typewriter is used, a character count will be required, and this can be particularly difficult with typewriters employing proportional spacing. If, on the other hand, a label starts immediately below the first scale point numeral on the x-axis, then it is easier to execute and both reading and writing logic are satisfied.

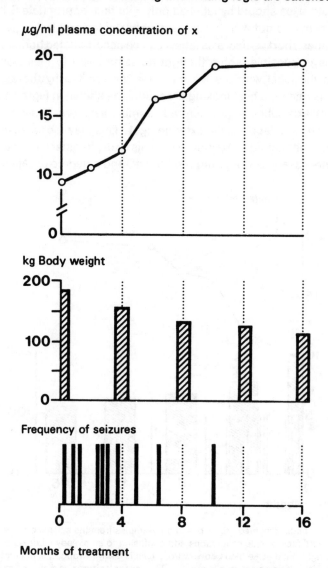

Figure 4.8B   Here the three sets of data shown in A have each been provided with their own scales, linked through time. This illustrates a scientific use of scales. Differently measured sets of data should never be superimposed on the *same* graph.

All labels should be in lower case, for the reasons given in Section 1.3.5. A capital may be used for the first letter of the first word only, never use a capital for the beginning of every word in a label or caption.

Standard abbreviations should be used for all units. The recommended procedure is to describe the variable and then follow this by the units in brackets. This system is simple and explicit, thus 'time (h)' or 'creatine

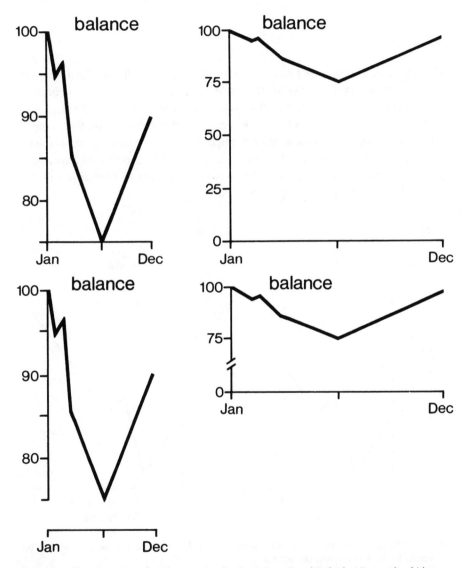

Figure 4.9 The distortion of scales can emphasise interesting details, but it can also hide or exaggerate the truth. It is honest to point out when a scale does not originate at zero.

clearance (µg/ml/min)'. Abbreviations of units should rot be pluralised. Grams is g and not gms; the latter would mean grams per metre per second. Similarly, abbreviations should not be followed by full stops unless these are part of the internationally agreed system. It is, however, common practice to insert full stops when using a patient's initials. If initials are used, it is advisable to follow them with the sex and age of the patient.

4.4.7. Keys

Wherever possible, direct labelling should be used on illustrations (Figure 4.12). This avoids the necessity for the 'double scanning' involved in finding the relevant part of the key and then referring back to the illustration.

If keys are used they should be positioned within the picture area if possible, and can be separated from the data by thin rules if necessary. The temptation to totally enclose the key in a box should be avoided. In the case of graphs, however, great care should be taken to ensure that key symbols are not likely to be confused with plotted points.

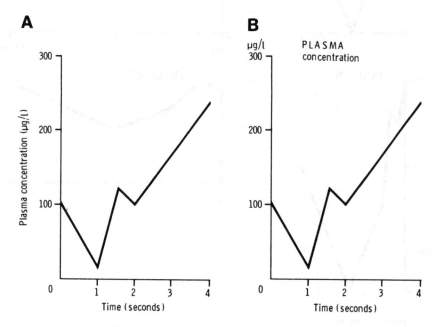

Figure 4.10 Comprehension is quicker when the amount of scanning required to find the essential features of the data is minimised.

B is simple and direct and leaves no doubt as to the subject matter. The labelling does not require turning of the head or the page.

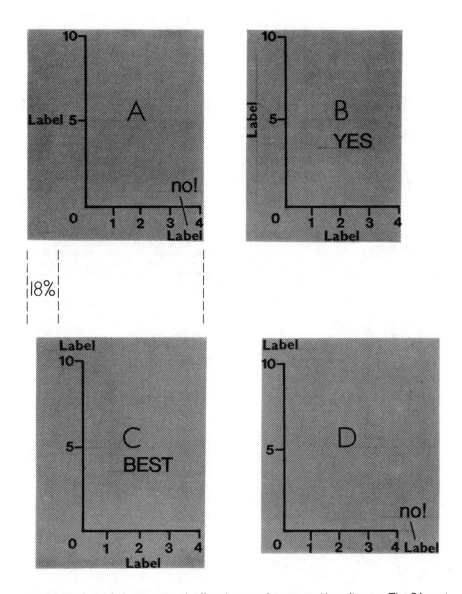

Figure 4.11 Axis labels can seriously affect the area of space used by a diagram. The *C* format is best for both slides and publication. (For slides the horizontal (landscape) orientation can also be used.)

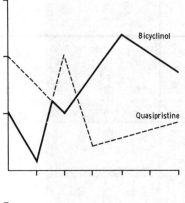
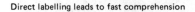

Direct labelling leads to fast comprehension

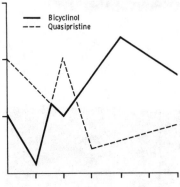

Indirect labelling requires double scanning, and comprehension is therefore slower. If, for some reason, a key is needed, then it is best shown within the graph area, as in this example.

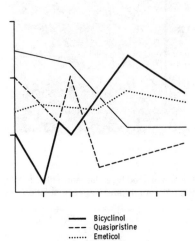

An undesirable situation may not be easy to avoid. In a complex graph, where space is limited, a separate key may become necessary.

Always try to avoid double scanning processes when planning data for a slide.

Figure 4.12 Direct and indirect labelling.

4.4.8. Headings

On illustrations for publication it is not usually necessary to include a heading or title in the artwork because the necessary information will be included in the figure caption. For slides, however, a heading may be necessary as a means of immediately identifying the subject. For dual-purpose illustrations it is often convenient to include a heading on the artwork, as this can easily be cut off subsequently if necessary. The content of headings for slides is discussed in Section 6.3.3.

5. Printed publications

5.1. Introduction

In the preparation of printed publications, the author will sometimes be responsible for the visual presentation of the text and any tables, as well as for the design and execution of the illustrations. This will be true in the case of journals produced from camera-ready copy and often for internally produced research reports. With camera-ready copy, however, the layout of the text will generally be specified by the publisher, and it is only in the case of reports that the author is likely to be involved in the design of the text. Factors to be considered in the visual presentation of text are dealt with in Chapter 2, and the more practical aspects of the preparation of the typescript are dealt with in Section 5.3.2. below.

Illustrations for books are usually prepared by the publisher, since they need to be of a very high standard and are sometimes in several colours. For journals and reports, illustrations need to be prepared quickly to save the author's time, yet with minimal sacrifice of quality. The methods recommended in this book are intended to satisfy these two requirements.

In some cases the author will wish to use an illustration as a lecture slide, as well as for publication in a journal. In this case it is essential that the artwork should meet the requirements for legibility in slide form, which will be more severe than those for the printed publication. If the standards recommended below for black-and-white artwork are applied, then the illustration will be legible as a slide and in print. This may mean, however, that the illustration must be considerably simplified in order to ensure its legibility as a slide, and it may be preferable to make a separate drawing for use as a slide, rather than oversimplify work required for printing. When planning artwork for publication, it is important to realise that it normally passes through three photographic stages as illustrated in Figure 5.1. This is why artwork should be prepared to a high quality, according to the principles and practices recommended in this book.

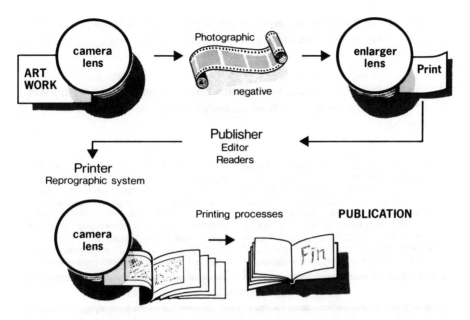

Figure 5.1 From artwork to publication. The artwork image passes through at least three photographic stages. Lens quality, focus and lighting can each adversely affect fine line elements in a drawing. This is why recommendations regarding minimum line thickness and good inkwork are important. (This illustration also shows how added interest can be given to a flow chart if pictorial elements and shading are used.)

5.2. Graphic standards for graphs, charts and diagrams in printed journals

5.2.1. Instructions to authors

All journals have 'requirements' or 'instructions to authors', and these are usually stated in each issue. When following these instructions, the author or artist should make sure that he is referring to a recent copy of the journal. Requirements vary somewhat and what follows is a synopsis of those features requested by the majority of journals.

Instructions to authors with respect to illustrations fall into two main classes. Some journals ask for lettering to be inserted on the artwork by the author 'in a size suitable for reproduction'. It is with this kind of instruction that we are mainly concerned here. Other journals state that lettering should be inserted on an 'overlay' (see Section 5.5 below). In this case the necessary legends will be set and inserted by the printer and the author's task is somewhat easier.

The size of the artwork required will depend on the page layout of the journal. There are two layouts in common use, as shown in Figure 5.2. These are:

1) the single-column layout in which the text is printed across the full width of the type area – this width will usually be between 15 and 20 cm or 6 to 8 in;

tempor cum soluta nobis eligend optio co impedit anim id quod maxim placeat face voluptas assumenda est, omnis dolor rep autem quinusd et aur office debit aut tue atib saepe eveniet ut er repudiand sint et recusand. Itaque earud rerum hic teneture au aut prefer endis dolorib asperiore repel tene sentntiam, quid est cur verear ne ad accommodare nost ros quos tu paulo ant fier ad augendas cum conscient to factor odioque civiuda. t tamen in busda tane est neque nonor imper ned libiding gen e cupiditat, quas nulla praid om umdant. It potius inflammad ut coercend magist and videantur. Invitat igitur vera ratio bene sa aequitated fidem. Imque hominy infant a cond que neg facile efficerd possit duo c effecerit, et opes vel fortunag vel ingen lif

the measure

tene sentntiam, quid est cur verear ne ad eam non possingdolor in reprehenderit in v accommodare nost ros quos tu paulo ante cum memorite itconsequat, vel lum dolor tum etia ergat. Nos amice et nebevol, olestias access potesteos et accusam et iusto od fier ad augendas cum conscient to factor tum poen legumlupatum delenit aigue duo est neque nonor imper ned libiding gen epular religuardoccaecat cupidtat non proi cupiditat, quas nulla praid om umdant. Improb pary minuitiqui officia deserunt mollits potius deserunt mut coercend magist and et dodecendense uptate velit esse molestait videantur. Invitat igitur vera ratio bene santos ad iustitiamieu fugiat nulla pariatur. Atl aequitated fidem. Teque hominy infant aut inuiste fact est dignissim qui blandit prae cond que neg facile efficerd possit duo conteud notiner sidolor et molestias exceptur

Print area

Figure 5.2 Illustrations may not normally exceed the type width (measure) or print area.

2) the double-column layout in which the type area is divided into two columns having an average width of about 7.5 cm (3 in) each.

A few journals have a three-column layout, but this is unusual. Whatever the page layout, the vital measurement is the column width or 'measure', as the illustrations must fit within this.

For economic reasons, most journals with a double-column format give preference to illustrations which will fit into one column when printed. As this is the most common journal layout, the requirements for illustrations to fit a 75 mm column width will be discussed in some detail. The effect of reducing an inappropriately designed table is shown in Figure 5.3.

5.2.2. Reproduction ratios

Limitations on manual dexterity will prevent the author/artist from preparing artwork of the same size as the final printed illustration. The artwork must therefore be prepared in a larger size and then reduced. As a general rule, work executed two times larger than its final size in print is usually convenient for the majority of simple illustrations such as charts and graphs. The artwork can then be made on an A4 sheet and contained within a standard rectangle of 13 × 20 cm (see 5.2.3). The relation between the size of the printed illustration and the size of the artwork is the reproduction ratio.

Printers, photographers, artists and authors will all need to give or receive instructions on reproduction ratios in the course of their work, and in the past this has often been a source of confusion. This has now been minimised by the introduction of an international standard whereby the linear reduction or enlargement is expressed as a percentage. The system operates as follows. An illustration which is to be reproduced the same size is reproduced at 100%. An illustration which is to be reduced to half its linear size is reproduced at 50%. An illustration which is to be enlarged to twice its linear size is reproduced at 200%. The formula is:

$$\frac{\text{A: desired final size (linear measure)}}{\text{B: size of artwork (linear measure)}} \times 100 = \text{per cent reproduction size}$$

Both A and B measurements are usually taken horizontally. 'A' is the type measure of either a single column or across two columns according to the printer's requirements, and 'B' is the horizontal dimension of the artwork.

The reader should not be confused by the fact that it is suggested elsewhere in this book that the longest dimension of the illustration is the most

A Appearance of small lymphocytes & transplantation immunity.

Species	Antibody Production		Transplantation Immunity	
	Days post coitus	Age equivalence	Days post coitus	Age equivalence
Man	112	0.78	98	0.60
Monkey	70	0.55	70	0.55
Cattle	118	1.00	264	3.52
Sheep	66	0.74	77	1.04
Pig	74	1.05		
Dog	56	2.45	40–48	0.33–1.38
Guinea pig	67	3.20	62	2.83
Opossum	28	0.58	24	0.43
Rabbit	31	0.36	27	0.18
Rat	28	0.58	24	0.22
Mouse	29	0.52	21	0.30
Hamster	26	1.1		
Chicken	24	0.40	23	0.36

After J. B. Solomon, 1971. Foetal and neonatal immunology.
North Holland Publishing Company. Amsterdam & London.

B |—— 30% ——|

Appearance of small lymphocytes & transplantation immunity

Species	Antibody Production		Transplantation Immunity	
	Days post coitus	Age equivalence	Days post coitus	Age equivalence
Man	112	0.78	98	0.60
Monkey	70	0.55	70	0.55
Cattle	118	1.00	264	3.52
Sheep	66	0.74	77	1.04
Pig	74	1.05		
Dog	56	2.45	40–48	0.33–1.38
Guinea pig	67	3.20	62	2.83
Opossum	28	0.58	24	0.43
Rabbit	31	0.36	27	0.18
Rat	28	0.50	21	0.22
Mouse	29	0.52	21	0.30
Hamster	26	1.1		
Chicken	24	0.40	23	0.36

After J. B. Solomon, 1971. Foetal and neonatal immunology.
North Holland Publishing Company. Amsterdam & London.

C Appearance of small lymphocytes & transplantation immunity.

Species	Antibody Production		Transplantation Immunity	
	Days post coitus	Age equivalence	Days post coitus	Age equivalence
Man	112	0.78	98	0.60
Monkey	70	0.55	70	0.55
Cattle	118	1.00	264	3.52
Sheep	66	0.74	77	1.04
Pig	74	1.05		
Dog	56	2.45	40–48	0.33–1.38
Guinea pig	67	3.20	62	2.83
Opossum	28	0.58	24	0.43
Rabbit	31	0.36	27	0.18
Rat	28	0.58	24	0.22
Mouse	29	0.52	21	0.30
Hamster	26	1.1		
Chicken	24	0.40	23	0.36

After J. B. Solomon, 1971. Foetal and neonatal immunology.
North Holland Publishing Company. Amsterdam & London.

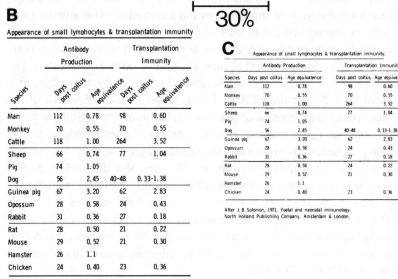

Figure 5.3 The design of tables affects the amount of space they occupy.
  A. The layout as originally typed. This will have to be reduced to fit a 70 mm column width (see C below).
  B. This layout is 30% narrower than A and fits a 70 mm column without further reduction.
  C. Layout A has been over-reduced to fit a 70 mm column. Many publishers would not accept this.
  Data supplied by kind permission of Professor I. Macintyre, RPMS London.

66

important, whether it be horizontal or vertical. This is true for illustrations in other media, but here we are concerned with requirements for printed publications. When an illustration intended for publication is likely to be reproduced in some other medium also, then the longest dimension will become significant (see Section 5.2.13).

5.2.3. Size of drafts and artwork

A4 is usually the most convenient paper size to work on, for the following reasons:

- Illustrations of this size can conveniently be sent through the post, without folding.
- Manuscripts are usually typed on A4 paper, and if the artwork is the same size this makes for convenience in handling.
- The proportions of an A4 page are roughly three units deep by two units wide, which corresponds with the standard slide format. This means that artwork for printed publications, if prepared in accordance with the guidelines set out below, will also be legible in slide form (see also Section 5.2.14 and Chapter 6).

Artwork will be accepted for publication if it is prepared within a rectangle measuring 20 cm × 13 cm (8 in × 5 in). A vertical format on an A4 page is preferred and when reduced by 50%, artwork of this size will fit comfortably within a 7.5 cm column width. It is essential, however, that the formulae given in Section 5.2.4 below should be used in respect of lettering, line thickness, line spacing and dot sizes if journal acceptance is to be ensured.

Throughout this chapter, our recommendations will assume that artwork is of the size recommended above.

5.2.4. Formulae

It will be found most convenient to make all measurements in millimetres.

A. Minimum capital letter height $= \dfrac{\text{overall long dimension of artwork}}{50\ (40)^*}$

B. Minimum line thickness $= \dfrac{\text{overall long dimension of artwork}}{600\ (500)^*}$

C. Minimum space between ruled lines = line thickness × 4

D. Minimum dot size = $\dfrac{\text{overall long dimension of artwork}}{500\ (400)^*}$

* For figures in brackets, see note in italic type below.

There are a few important points to emphasize regarding the use of these formulae.

In A – the capital letter height of lettering is used as a convenient form of measurement. This does *not* mean that the use of capital letters is recommended. Quite the reverse is true for reasons given in Section 1.3.5.

Most journals require that the capital letter height of lettering in the final printed form should not be less than 2 mm. If the formula above is applied to the 13 × 20 cm artwork area, this gives a capital letter height of not less than 4 mm. This size will be acceptable whether a vertical or horizontal format is used.

In B – the minimum line thickness for artwork prepared in the 13 × 20 cm artwork area is 0.35 mm using the 'micronorm' standard for technical pens (Figure 5.4).

In C – the minimum space between lines refers particularly to the use of fine parallel lines such as might be used for 'shading' in bar charts.

In practice it has been found that these formulae can be successfully applied to artwork required for both publication or slides and in either the vertical or horizontal formats. The figures enclosed in brackets make for a slightly bolder drawing, and this might be preferred for multi purpose artwork.

5.2.5. Character style

Where the publisher requires the author to insert the lettering on artwork, it is best to use a simple undecorated letter form. Micronorm stencilled lettering or IBM Directory typescript will be suitable. If dry-transfer lettering is used, a simple face such as Univers or Helvetica is preferable.

5.2.6. Character spacing (using stencilled lettering or dry-transfer letters)

For most styles of lettering, the correct inter-letter space will be approximately the same as the stroke width or line thickness of the characters. The

68

Actual size

mm
0·35

───────────────── 0·5

───────────────── 0·7

───────────────── 1·0

Reproduction at 50%

mm
───────────── 0·35

───────────── 0·5

───────────── 0·7

───────────── 1·0

ISO micronorm line weights
recommended for artwork with
an area of 13 x 20 cm

Black and white tones reproducing as grey

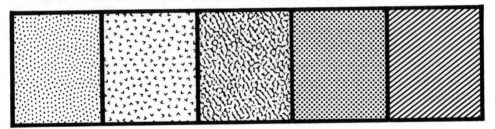

Actual size as recommended for artwork

Reproduction at 50%

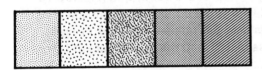

Figure 5.4 The line weights and pattern densities recommended here will help to ensure that artwork reproduces well for either publications or slides, provided the artwork area does not exceed 13 × 20 cm (5 × 8 in).

correct word spacing will be the width of the letter 'e'. In the case of stencilled lettering, the stencils are usually designed in such a way as to facilitate correct spacing (see Section 12.2.5). Further comments on spacing in relation to slides are made in Section 6.4.7.

Capital letters are much more difficult than lower case letters to space correctly (see Figure 5.5) and this is another reason for restricting their use. A next to W can be particularly difficult to space satisfactorily, and so can L next to any other letter. The space on the left of an L will often look much less than the space on the right. A certain amount of aesthetic judgement is therefore required in positioning the letters so that they appear to be evenly spaced. It must also be remembered that letters touching each other are likely to be illegible and confusing and letters too far apart will slow down the reading process. It is therefore important to become aware of the appearance of good spacing by studying well-designed and printed materials.

5.2.7. Space between lines of text

Sufficient space must be left between lines of lettering so that the descenders and ascenders of letters in adjacent lines do not touch. When using Helvetica, Univers, DIN and most other modern sans serif styles, the total amount of space occupied by each line, as measured from the base of the x-height on one line to the base of the x-height on the next, should be not less than one-and-a-half times the capital letter height. Thus if a 4 mm capital letter height is used, the distance between lines should be not less than 6 mm (Figure 5.5).

5.2.8. Tones and shading

In some cases tones or shading may be required. These must meet certain standards in order to ensure that they will reproduce well in print (Figure 5.4). For tones made up of dots, a suitable dot size is 0.5 mm, with a minimum of 0.5 mm between dots. For shadings built up from parallel lines or cross-hatching, the minimum line thickness of 0.35 mm, with a minimum of 1 mm space between lines, will be suitable.

It is also worth bearing in mind that some kinds of shading will reproduce better than others when copied electrostatically. Large dark areas generally do not copy well and shading made up of discrete points will therefore copy better than a continuous tone.

If several tones are needed, they should be easily distinguishable. Subtle

70

distinctions may be lost in copying or microfilming. Ideally, shading should be avoided altogether in publications which will be microfilmed. The way tones are used requires careful thought. Some common errors are illustrated in Figure 5.6.

Any lettering superimposed on a tone should be placed on a white sur-

The *linear* space between the characters is equal, but the effect uneven.

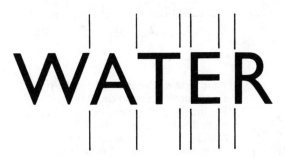

Here the *area* of space between characters has been judged by eye to be approximately equal; the spacing now appears to be more even.

The space below and above must allow for descenders/ascenders

The distance at which lines of lettering are to be viewed affects the amount of space required between lines. The example above shows close spacing which is suitable for text to be read 'in the hand'. More space between lines would be needed for a positive slide and even more for a negative slide.

The amount of white space between lines of type should always be greater than the space between adjacent words.

Figure 5.5 Spacing between characters and between lines.

round (Section 12.5.3). Some background patterns can severely reduce the legibility of superimposed lettering, and coarse, regular dot patterns are among the worst offenders in this respect (Spencer et al., 1977).

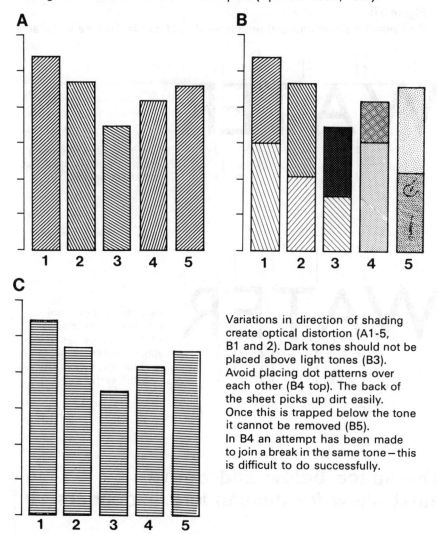

Variations in direction of shading create optical distortion (A1-5, B1 and 2). Dark tones should not be placed above light tones (B3). Avoid placing dot patterns over each other (B4 top). The back of the sheet picks up dirt easily. Once this is trapped below the tone it cannot be removed (B5). In B4 an attempt has been made to join a break in the same tone – this is difficult to do successfully.

Avoid horizontal tones. These often create thicker lines at the top or bottom of columns.
An example of inaccurate cutting is shown in C5.

Figure 5.6 Tones – some problems.

5.2.9. Positive versus negative images

Although negative images (white type on a dark background) have a strength which can sometimes be used to good effect in print, their execution requires special care.

Tinker (1963) has shown that continuous text printed in white on black is less legible than the same text printed in black on white. This is because it is a characteristic of the human visual system that bright images seen against a dark background always appear slightly blurred and thickened. There is therefore a tendency for letters to appear to fill in and run together, and this must be taken into account in designing illustrations in negative form. It is advisable to slightly overspace any lettering or symbols, and if possible to use characters which are slightly larger than the recommended minimum size.

These precautions are especially important for illustrations which are intended for use as slides as well as in print, since the projection of the image will accentuate the blurring and thickening effects. Where a positive illustration is also intended to serve as a master for the production of a negative microfilm image, the same measures are advisable.

5.2.10. Contrast

A high level of contrast between ink and paper is essential if a drawing is to survive the various reproduction processes to which it will be subjected without loss of quality. This means that the paper must be as white as possible and the ink as black as possible. Many papers are not white, and many so-called black inks are dark blue rather than black. Tracing paper and typing papers should therefore be avoided, as should all ballpoint and felt-tipped pens, because these rarely if ever produce a black image or a sharp-edged line. Suitable materials are discussed in Chapter 11.

5.2.11. Oversize artwork for publication

Sometimes the nature of the information is such that it is not possible to accommodate it within the standard 20 × 13 cm rectangle. There are ways of dealing with oversize material, but it is still important to observe the publisher's requirements to ensure legibility and acceptance of the data by the journal.

For moderately oversize artwork, an appeal can be made to the editor to allow the illustration to be printed across two columns. This procedure allows the horizontal measurement of the original to be doubled, without affecting standards for line thickness or character size. If this procedure is adopted, it is important to note on the original or its cover sheet that the illustration is 'designed to print across two columns'.

Even larger originals can be prepared to fit within the overall print area of the page. If the illustration has a horizontal orientation, then this will mean that the reader must turn the journal sideways in order to study the data. This option allows for originals which may have a long axis of 50 cm (20 in), and the standards for line thickness and letter height will again remain unchanged.

Sometimes the original cannot even be fitted into an area which is twice the printed page size. This often happens with long-duration studies where a patient's progress is recorded month by month or year by year. If such data are to be published, then the author may have to edit the information drastically, leaving only the bare essentials. In some cases, however, the point of the illustration may be lost if the data are edited. It may then be necessary to split the illustration into two or more parts in order to show the detail. Whichever alternative is adopted, it is essential to observe the standards governing size of lettering, or the printed version will not be legible.

It must be emphasised that wherever exceptions to the standard 20 × 13 cm image area are planned, the type area available must be precisely measured and the advice of the journal editor sought before work begins.

5.2.12. Compatibility with standards for slides

In the majority of cases, artwork with an image area of 20 × 13 cm which has been prepared in accordance with the standards recommended above will be suitable for the production of standard 35 mm slides. If, however, it is intended that artwork should be used for a slide as well as for a printed illustration, there are certain characteristics of the photographic methods used in making slides which must be borne in mind. When black-and-white slides are made from black-and-white artwork, a high-contrast film is often used. As a result of this, large areas of black or very thin and very thick lines on the same drawing will not reproduce well. The line thickness increment from 0.35 mm to 1.00 mm can be tolerated, but further differences in size may not be acceptable. This limitation may therefore need to be taken into account in preparing dual-purpose artwork, though it is always as well to check with the photographer concerned. This difficulty does not arise with printed illustrations because different reproduction techniques are used. It must always be remembered that the artwork is only the starting point for a scientific illustration, and the effects of subsequent reproduction processes must be taken into account in preparing it.

Non-standard or oversize artwork may be unsuitable for reproduction in slide form. If an illustration is designed to print across two columns, for example, it is likely that the lettering will be too small to be legible on a slide. Similarly, if the proportions of the artwork are not 2:3, the reproduction ratio required to fit the illustration onto a slide may reduce the height of the lettering excessively.

If oversize artwork has to perform a dual function then the formulae in 6.2.4 must be applied in legibility is to be ensured.

5.3. Guide lines for camera-ready copy

5.3.1. Copy for journals or proceedings of meetings

It is becoming increasingly common for journals to ask their contributors to submit papers in the form of camera-ready copy. The typewritten pages are then photographed and printed by offset lithography, thus eliminating expensive typesetting. This does mean, however, that the manuscript and illustrations must be prepared very carefully.

Journals produced by this method usually issue authors with detailed instructions in relation to the preparation of the typescript and illustrations. Requirements vary from journal to journal so it is difficult to make any general recommendations here. The use of bond paper (see Section 11.2.1) and a carbon ribbon will almost always be specified for the text, in order to give maximum contrast. The page size and type area will also be specified, and these instructions must be followed precisely. In many cases the typed pages will be reduced in size before printing, so that the final printed page may be reproduced at 75% of the original. This will have implications for the size of lettering on the illustrations, and the author must therefore pay particular attention to any statements in the instructions relating to reduction ratios.

The position of any illustrations must be carefully planned in relation to a draft manuscript, and in the final typing, spaces of the appropriate size must be left for them. The finished illustrations must fit these spaces exactly.

If the formulae in 5.2.3 are employed then the bracketed numbers *must* be employed when making illustrations for camera-ready copy. The illustrator must bear in mind that the original artwork will be reduced approximately 50% to fit the space left for it in the typed text. The whole of the page will then again be reduced to 75% before it appears in the final form.

5.3.2. Copy for reports

Many reports, both published and unpublished, are also produced by photo-graphing a typewritten manuscript and printing by offset lithography. A4 is a popular format for reports, and in this case it is usual to use an A4 format for the manuscript and to photograph it without any reduction in size. The layout and design of reports via camera-ready copy will depend on the sort of information being presented. If the report is to be merely read once or twice then possibly discarded, ease of reading is the prime consideration.

For reasons given in Section 2.2.2 paragraphs flush left and separated by a space from the preceding one create a more readable page than one in which paragraphs are only indented. Suggestions for the layout of headings are given in Section 2.2.1 and in Figure 2.1. It is important *not* to put headings of scientific reports in capitals as this nullifies the special or symbolic use that certain capitals may have and slows up the reception of information that is possible with rapid scanning. For example:

1) A.C.T.H. IN SHONA BLOODGROUPS – A W.H.O. REPORT.
2) ACTH in Shona blood groups – a WHO report.

In example 1, there is no differentation in the use of capitals. In example 2, capitals are used without full points to denote (a) a drug, (b) an international organisation, and at the beginning of a proper name for an ethnic population. If, therefore, headings in reports are separated from the text by appropriate spaces, ruled lines, or different type styles, this is preferred to using capitals indiscriminately.

If the report is to be used as a reference document, the suggestions given in Section 2.3 in relation to bibliography design should be followed.

Reports to be printed in this way from camera-ready copy should be typed on good quality bond paper with a carbon typewriter ribbon. The type area should be bounded by relatively generous margins, which should be used consistently. If the report is to be printed on one side of the paper only and is to be bound with a plastic spiral binding or a slide binder, then a suitable left-hand margin must be allowed for. If the report is to be printed on both sides of the paper, then the position of the type area must be adjusted for left-hand pages. If some other binding method is used such that the inner margin is not encroached upon, then the type area can be moved towards the spine, thus increasing the outer margin slightly. As a general rule there will be no advantage in increasing the line length. Margin widths totalling 8 cm (3 in) will result in a line length of about 60 characters and any further increase in length may reduce legibility, particularly if the lines are single spaced.

As in the case of camera-ready copy for journals, the position of any illustrations must be carefully planned in advance, and spaces must be left in the typescript to accomodate them. In the case of full-page illustrations it will sometimes be possible to prepare the artwork within the A4 format, thus avoiding any need for subsequent reduction. Whether or not this is possible will depend on the amount of detail in the illustration. Smaller illustrations may need to be prepared at twice their final size however. When the artwork is same-size, the lettering can be typewritten on the same typewriter as the text, but if some illustrations need to be reduced it is better to use stencilled lettering so that its style and size is consistent throughout.

5.4. Original artwork versus copies

5.4.1. Reasons for copying artwork

Most publishers require 'two or three copies' of illustrations. One of these is for the printer, and others may be for the readers or editors. Xerox or similar copies of black-and-white illustrations will be adequate for readers and editors, since they will be interested in the content rather than in the quality. The printer, however, will require the original or a good quality copy.

There is no doubt that the best results in print are obtained directly from high-quality originals, but there are several reasons why the original should remain in the protection of the author:

– lost or damaged copies can be replaced if the author retains the original;
– it may be important for the author/artist to retain the original if the copyright is also retained;
– the original artwork may be required for additions or for adaptation to other media.

If the original is to be kept as a 'file copy' for any of the above reasons, then good quality copies will be essential.

5.4.2. Copying methods

Thermographic or electrostatic (Xerox) processes rarely if ever produce copies of good enough quality to replace the original for printing purposes. What is needed is a true facsimile copy. Traditional photographic processes produce true facsimiles via a negative. Negatives are a useful source of further copies if these are required, but traditional photography is time consuming and there are quicker ways of making facsimile prints.

The quickest and cheapest alternative is the diffusion-dye-transfer method, often called reflex copying or PMT (photo-mechanical transfer). The two most efficient systems have been developed by Agfa Gevaert under the names of 'Copyrapid' and 'Copyproof'. Both of these systems produce direct-contact reflex prints using special materials and a simple 'semi-wet' processing unit. The prints take only a few seconds to produce and they dry in room conditions in about 15 minutes.

Both Copyrapid and Copyproof systems enhance contrast very strongly. Providing that the processing exposure is correctly set, 'dirty' backgrounds will appear white, and pale grey lines will appear black. This effect is extremely useful when relatively low-contrast originals such as ECG, EEG or UV traces are used. The copies will therefore be of better quality than the artwork in some cases, and they are of such a high standard that they will always be accepted for publication. In the case of the simpler Copyrapid system, however, it is important that cut-and-stick corrections should be done carefully and the use of tapes kept to a minimum if best results are to be achieved.

With the Copyrapid system it is only possible to produce same-size copies. This is a further reason for restricting the size of the artwork to A4. With the Copyproof system, however, it is possible to reduce or enlarge the size of the copy via a 'process' camera. This is particularly useful for oversize drawings. If a reduced copy is taken, correct sized lettering can then be stripped-in using the cut-and-stick method (see Section 12.3.2) thus ensuring legibility. A final copy can then be taken after the lettering has been added. Both Copyrapid and Copyproof systems make it possible to take a same-size copy of the original before and after lettering has been applied. This is useful where a journal requires the illustrations without lettering, but the same artwork is to be used to make a slide which does require lettering.

5.5. Overlays and cover sheets

These perform two essential functions. Firstly they protect the surface of the drawing or copy from damage, and secondly they can be used to make notes to the editor or printer. The position of any lettering or other printed items to be inserted on the drawing will also need to be indicated on the overlay.

The overlay should be attached to the illustration in the manner shown in Figure 5.7. This will ensure that it will not move, and that the position of any printed items to be added is indicated accurately. For line illustrations the

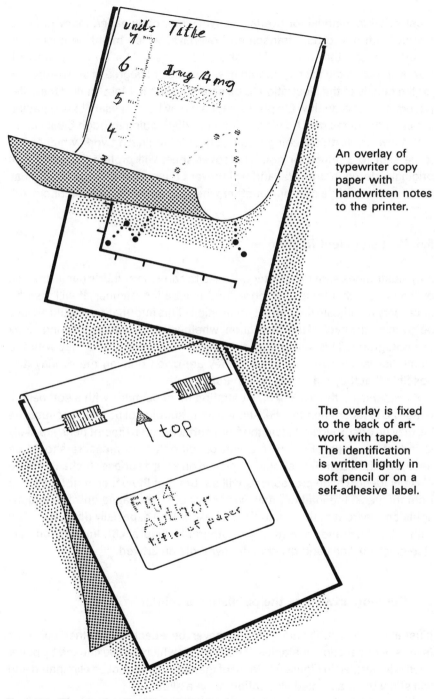

An overlay of typewriter copy paper with handwritten notes to the printer.

The overlay is fixed to the back of art-work with tape. The identification is written lightly in soft pencil or on a self-adhesive label.

Figure 5.7   The use of overlays to protect artwork.

most suitable material for overlays is typewriter copy paper. Many journals instruct authors to use 'transparent' overlays, which might be misunderstood to mean that a clear film should be used. The only requirement, however, is that the overlay should have a sufficient degree of show-through for the details of the illustration to be seen through it. Most 'bank' typewriter papers have this quality. On photographs it may be necessary to use tracing paper as the detail may be more difficult to distinguish through bank paper.

It is always worth attaching a cover sheet to original artwork, whether it is to be sent to a printer or not. The cover sheet will protect the surface of originals stored in drawers, and will prevent the sort of damage which may occur as items slide over one another when the drawer is opened or closed.

5.6. Marks of identification

Any illustrations sent to a printer will need to carry the author's name, the title of the paper, chapter and/or book, and illustration number. It will also be necessary to indicate the correct orientation. This information should always be given on the back of the illustration, whether it is an original drawing, copy or photograph. This will ensure that the identification marks stay with the illustration at all times, whereas if they are given only on the overlay it is possible that they may become separated.

Any marks on the back of an illustration must be made with a soft pencil. Grade 2B is recommended. Felt-tipped pens should not be used because the ink may be absorbed and stain the front surface, and ballpoint pens are likely to indent the paper. Even with a soft pencil only light pressure should be used, and the illustration should be placed on a hard surface. A sheet of plate glass is ideal. These precautions will ensure that the paper is not indented from the back. Indentation, if allowed to occur, can cause unwanted highlights on the front of the illustration. These will be especially difficult to deal with if the illustration is a glossy photograph (Figure 5.8). Indentation may also crack ink lines and dry-transfer symbols on artwork.

5.7. Sending artwork to the publisher or printer

Illustrations for publication should never be attached to other items by means of paper clips or staples. The damage which can be caused by paper clips is illustrated in Figure 5.8. Most of these marks cannot be eliminated and will show up in a printed illustration or on a slide.

The artwork should always be posted flat, and never folded or rolled.

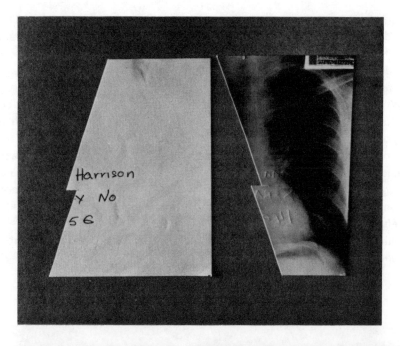

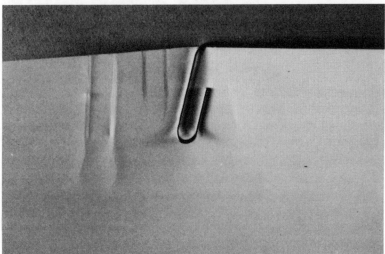

Figure 5.8
  Top. Damage caused by hard points on the back of a photograph may be seen on the front.
  Bottom. Damage caused by paper clips is usually irreparable.

Rolling causes stress on the surface of the drawing which can crack ink lines and dry-transfer products such as lettering or tones. Any creases arising from folding or rolling will be irreparable. If artwork is prepared in the recommended A4 format, however, it can be posted flat without difficulty and with the minimum risk of damage.

When illustrations are sent by post, they should be packed between two sheets of stiff card. If they vary in size, the smaller ones should be lightly fixed with drafting tape to sheets of thin card which have been cut to fit neatly into the envelope. This will prevent the illustrations from sliding about inside the envelope. For the same reason, the envelope itself should be just large enough to accept all the illustrations. If it is too large the items will move inside it, possibly causing damage and sometimes even breaking open the envelope.

Valuable illustrations should be sent by registered post or recorded delivery.

6. Slides

6.1. Introduction

Slide presentations have become an important medium for the exchange of ideas and data among research workers and others, but in spite of their extensive use, their effectiveness in conveying information is often much less than it might be. There are several factors which may contribute to this lack of effectiveness:

a) uncomfortable seating
b) external distractions
c) poor acoustics
d) poor projection facilities
e) poor delivery on the part of the lecturer
f) slides which are incomprehensible or illegible.

The lecturer normally has no control over the first four items, so his delivery and the quality of the slides are of paramount importance.

Any slide presentation must be very thoroughly planned in advance. The content of the slides, their sequence and the total number required must be carefully thought out in relation to the subject matter, the audience's familiarity with it, and their capacity for absorbing information in a lecture situation. *The design of the slides themselves must take into account the limitations of the medium for conveying detailed information*, and above all the data must be legible.

In presenting information on slides, it is important to realise that there is a fundamental difference between looking at information on a printed page and looking at a slide. When looking at a printed page, the reader is self-paced. He can study the information for as long as he wishes. When looking at a slide, he is limited to the amount of time that the lecturer allows him. It is therefore essential that a slide should contain only as much information as the audience can reasonably be expected to assimilate in the time the speaker allows. Furthermore, if the information is to be understood rapidly, it must be presented in a visually direct and simple form. Several simple slides, each making perhaps one to two 'statements' and projected for a short time, will be much more effective than a complex slide shown for a longer period.

It is also essential to pay adequate attention to the question of legibility. Most slides will be legible if viewed at close quarters in a seminar situation, but if they are not prepared according to the standards recommended in this Chapter, they are likely to be legible in only the first five rows of a large lecture hall. It is an insult to the audience if data are presented in an illegible form. Some lecturers will argue that it is immaterial if the audience cannot read the captions on a graph, so long as they can see the lines linking the data points. Even if this were true, it is still extremely frustrating and irritating for an audience to be presented with information which cannot be read.

6.2. Planning a slide presentation

6.2.1. The use of 'introduction' and 'conclusion' slides

A lecture or talk accompanied by slides should be planned in the same way as a written exposition of the subject, to the extent that it should have a clearly defined structure consisting of a 'beginning', a 'middle' and an 'end'. It is now common practice to use one or more 'introduction' slides, then the slides illustrating the body of the talk, and finally one or more 'conclusion' slides.

In planning a presentation of a relatively general nature, it is often helpful to begin with an introductory slide which lists all the main topics to be covered. This will enable the audience to understand the logical structure of the talk that they are about to hear. Similarly, a final slide or slides summarising the main conclusions of the talk is likely to clarify the issues described and to help the audience to remember the salient points.

In a shorter presentation describing a specific piece of experimental work, an introductory slide might be used to make a statement of the problem. This should be done in as interesting a manner as possible, perhaps by asking a question. Further introductory slides might summarise what was previously known about the subject. The methods used and the results obtained would then be described briefly in the body of the presentation. The conclusion slides should be worded in such a way as to stimulate further interest in the subject, for example by summarising the questions which remain un-answered, any new questions raised by the results, and possible future lines of research.

6.2.2. The body of the presentation

The planning of the middle section of the talk will depend to some extent on the time available. In a ten minute presentation of an experimental study, for example, there is very little time for expanding on details and the results must therefore be described very simply. It is important to isolate the most significant aspects of the work, and then to present them as dramatically as possible; in presenting the results of a clinical investigation the details of the condition of the patient before and after treatment could be shown in a striking manner. Drama in slides is not only permissible, but often highly desirable – providing, of course, that the bounds of scientific accuracy are not exceeded.

The number of slides required for this part of the presentation will depend on the subject matter, the style of lecturing adopted and the length of time available. It is important to have enough slides to illustrate the subject adequately, bearing in mind that the audience will not be so familiar with the subject as the lecturer and his close associates. It is very easy to take for granted a level of knowledge on the part of the audience which does not in fact exist. On the other hand, the number of slides should not be so great that the lecturer is forced to rush through them in order to complete the presentation within the alotted time. Again it is easy for a lecturer who is familiar with his subject to underestimate the time required by the audience to assimilate each slide.

In terms of style of lecturing, there are two approaches to the use of slides which can be used either separately or together in the same lecture. A spoken presentation can be very effectively complemented by the use of a series of very simple slides which reinforce what the lecturer is saying. These may contain only a few words or a very simple illustration, but they can be used to build up more complex concepts. Such slides can be shown without any interruption in the lecturer's argument, and indeed they can often be used as prompts if necessary. This system contrasts with the more traditional approach of using fewer, more complex slides which require a deliberate pause for explanation.

Where a large number of slides is used while the lecturer is talking, they should carry only *one* simple message each, and their timing should be very carefully coordinated with the spoken presentation of the same messages. In this way it would be theoretically possible to use as many as thirty slides in a ten minute talk. This would allow twenty seconds per slide. This approach will not necessarily be suitable for every situation and if the alternative technique of halting to explain each slide is adopted, this may limit the number of slides which can be shown in a ten minute lecture to eight or ten.

Sometimes during a lecture, particularly if the more traditional approach is adopted, there will be periods in which there are no appropriate slides to be shown. In this case, the previous slide should not be left on the screen or the audience will be studying this instead of listening to what is being said. The same will happen if the lecturer prematurely advances to the next slide. The solution to this problem is to insert blank slides into the sequence at the appropriate points. Under no circumstances should a slide containing any kind of visual message be used as a blank, no matter how unrelated that message is to the subject of the talk. Patterns or landscape views, for example, will merely provide another source of distraction. When it is necessary or desirable to refer to a slide more than once during a lecture, duplicates should be used.

6.2.3. Constructing a story-board

A convenient way of planning a slide presentation is to use a story-board. This can be constructed in the following way:

1) Prepare a plywood board about 70 cm (27 in) wide by 50 cm (20 in) deep.
2) Place small pins or nails at regular 5.5 cm ($2\frac{1}{4}$ in) intervals across the board in rows 7.5 cm (3 in) apart.
3) Make a number of 3 × 5 cm (2 × 3 in) cards and punch holes in one of the shorter edges so that the cards will hang vertically from the pins on the board.
4) Divide each card in half horizontally using a rule.

Each card represents one slide. The top half can be used for a synopsis of the spoken commentary which accompanies the slide, and the bottom half for the wording on the slide or a rough sketch if it is an illustration. The board will take forty cards, which should be enough to plan the major sections of a lecture. The sequence of the cards can be changed very easily, and cards can be added or discarded at will. Planning a lecture in this way can often avoid the preparation of slides which will be discarded as unnecessary later on.

6.2.4. Rehearsal

The ultimate test of any slide presentation is to try it out on an audience. Ideally all slide presentations should be rehearsed, even if only in front of colleagues who may be familiar with the subject. They will probably vary in the speed at which they are able to assimilate visually presented data and this, in itself, will be helpful. They are also likely to be able to point out any obvious defects in the content and sequence of the slides which the lecturer himself may not be aware of.

6.3. The content of slides

6.3.1. Word slides

Introduction and conclusion slides will usually carry a simple message in words rather than an illustration. This will also be true of some of the slides used in the body of the presentation. If these slides are to be used to complement the spoken word, rather than as exhibits in their own right, then they must be simple. The audience cannot be expected – and will not be able – to read a complex slide and attend to what the lecturer is saying at the same time.

The ideal maximum number of words for these slides is about thirty (Figure 6.1). They must therefore be very carefully worded in order to make the best use of the space available. This can be done by means of the 'telegram' technique which relies on the judicious use of keywords and linking words. The interpretive effort required on the part of the reader means that he is more likely to remember the content as a result of having had to make a contribution from his own knowledge and experience. He is not expected to make this interpretive effort unaided, however, since the lecturer will be there to give guidance. All slides shown in a teaching situation should require the presence of the lecturer, rather than being self-sufficient. Slides differ fundamentally in this respect from illustrations intended for publication, which should be largely self-explanatory.

6.3.2. Data slides

In describing research methods and results on slides, it is more important to stimulate interest than to give detail. In tables and illustrations intended for printed publications it is often desirable to give as much detail as possible. With slides, however, there is a limit to the amount of information which can be displayed without loss of legibility and to the amount which can be assimilated by the audience in the short time available. The lecturer must therefore ask himself what point he is trying to make on each slide, and include only the information which helps to make this point. The selected information must then be presented as concisely and simply as possible.

Tables designed for slides must be kept extremely simple. Decimals should be rounded as much as possible – ideally to the nearest whole number – and fractions should be omitted. Similarly, standard errors should be limited or omitted. It must be remembered that the function of a table in

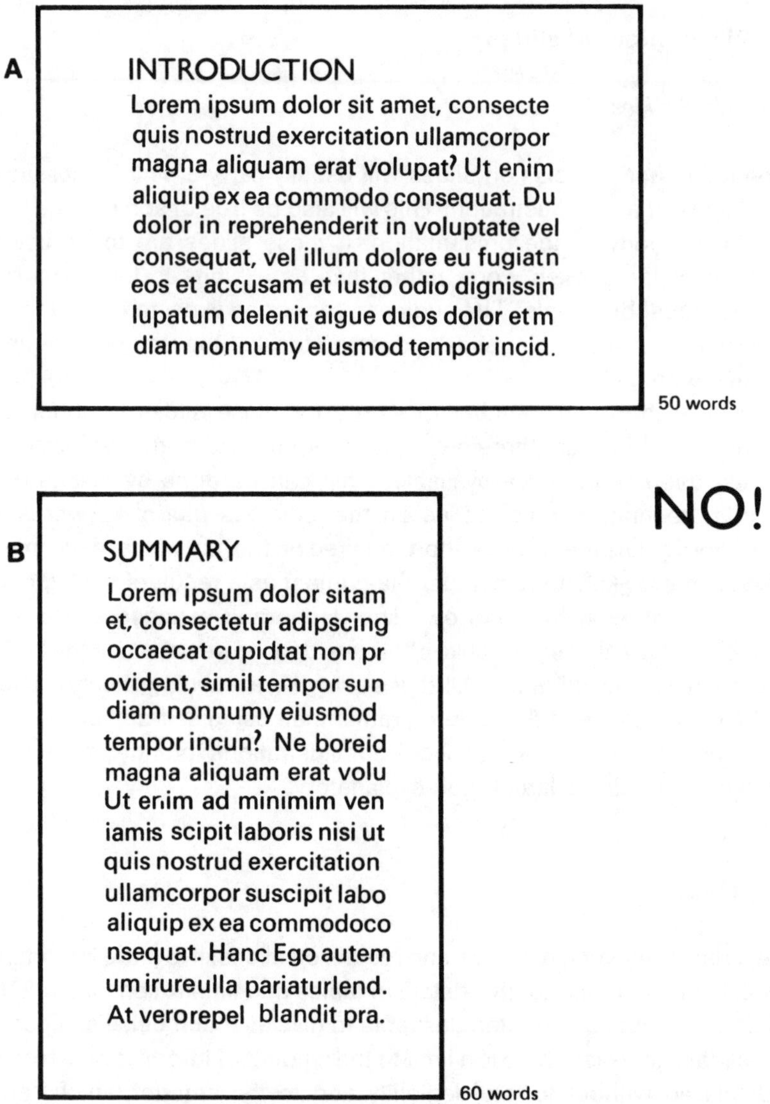

A

INTRODUCTION

Lorem ipsum dolor sit amet, consecte quis nostrud exercitation ullamcorpor magna aliquam erat volupat? Ut enim aliquip ex ea commodo consequat. Du dolor in reprehenderit in voluptate vel consequat, vel illum dolore eu fugiatn eos et accusam et iusto odio dignissin lupatum delenit aigue duos dolor et m diam nonnumy eiusmod tempor incid.

50 words

NO!

B

SUMMARY

Lorem ipsum dolor sitam et, consectetur adipscing occaecat cupidtat non pr ovident, simil tempor sun diam nonnumy eiusmod tempor incun? Ne boreid magna aliquam erat volu Ut enim ad minimim ven iamis scipit laboris nisi ut quis nostrud exercitation ullamcorpor suscipit labo aliquip ex ea commodoco nsequat. Hanc Ego autem um irureulla pariaturlend. At verorepel blandit pra.

60 words

Figure 6.1A, B   Do NOT try and fill the screen with words, even if they are legible according to the criteria given in this book. The layouts above will have an immediate and lasting 'switching off' effect on an audience. The slide should never be confused with the written paper.

slide form is not to give a definitive statement of the results of an experiment. This is the function of a written paper. The slide should be designed to show the important trends clearly and simply.

With graphs and charts too, the aim should be to demonstrate trends rather than to give detail. Did the temperature go up or down, and if so, how fast? If

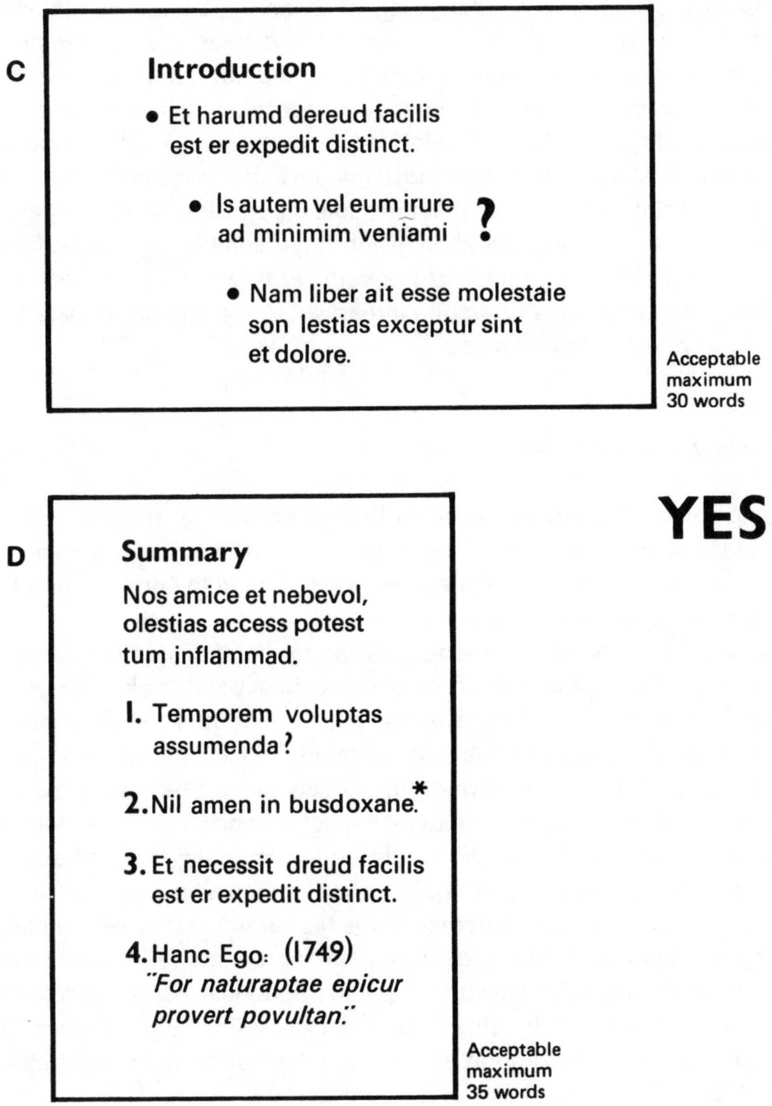

C

Introduction

- Et harumd dereud facilis est er expedit distinct.

    - Is autem vel eum irure ad minimim veniami **?**

        - Nam liber ait esse molestaie son lestias exceptur sint et dolore.

Acceptable maximum 30 words

YES

D

Summary

Nos amice et nebevol, olestias access potest tum inflammad.

1. Temporem voluptas assumenda?

2. Nil amen in busdoxane.*

3. Et necessit dreud facilis est er expedit distinct.

4. Hanc Ego: (1749) "For naturaptae epicur provert povultan."

Acceptable maximum 35 words

Figure 6.1C,D Examples of a distinctly more appealing layout for word slides. Typographic techniques have been employed to group statements, to emphasise a question or identify a quote. These variants can help to stimulate interest in the subject.

these basic facts are obscured by statements regarding error and probability, the essential message may be missed. Given the limited space available on the screen, slides are best used for the visual presentation of concepts which would be otherwise hard to grasp. Thus, although degrees of accuracy in scientific data are obviously very important, they are better spoken by the lecturer or provided in a handout than given on the slide. The statement '81.20% ($\pm$ 0.03) success rate' may have the advantage of accuracy, but on the screen the message would be communicated more rapidly by wording such as 'success over 80%'. There may be occasions when the statistical significance of the results is of major concern in the presentation, but even so, the data must not be overcrowded. It may be better to illustrate the results first in a very simple manner so that the audience can appreciate the general trends, and then to present a slide giving more complex and scientifically accurate data. There is a limit to the complexity which can be achieved however, and in the case of graphs in particular, the number of data lines should be limited to four at most.

6.3.3. Headings for data slides

Headings on data slides are useful for the lecturer as a means of identifying the slide, and for the audience as an indication of the content. Headings should be placed at the top of the slide rather than at the bottom, and they should be ranged left rather than centred.

The main function of the heading as far as the audience is concerned should be to stimulate interest. It need not be totally self-explanatory, since the lecturer will be explaining what the slide is about. The most effective headings, therefore, will be concise and simple. For example, a heading such as 'The infra structure of polymorphic G cells in chickens with diabetes mellitus' would have more immediate impact if shortened to 'G cells and diabetes – chickens'. Better still would be an intriguing heading such as 'G cells – do they exist?'. The audience will immediately begin to wonder what the answer might be, whereas in the first example they would still be reading the heading. Sometimes it may be appropriate to include a short reference, particularly if the lecturer is giving a historical resume. In this case the heading might be 'The structure of G cells – Fox 1921'. Typographic variations can be used to good effect in this type of heading, as shown in Figure 6.2.

A

THE INFRA STRUCTURE OF POLYMORPHIC G-CELLS IN CHICKENS WITH DIABETES MELLITUS

B

G-cells and diabetes - Chickens

C

G-CELLS "Do they exist?"

Figure 6.2
   A. The subject heading has been typed in capitals. Many typewriters produce poorly spaced capitals and this makes reading slower. Apart from this, too many words create a 'heavy' impression.
   B. This version is short and to the point, yet retains the essential message.
   C. This is even shorter, and the use of different type styles adds interest. Most important of all, however, is the challenging question. The audience is immediately interested.

6.4. Graphic standards for slides

6.4.1. Originals

Tables, graphs, charts and diagrams photographed from printed books, reports and papers generally do not make good slides for two main reasons. Firstly, printed information is usually presented in too complex a manner and contains far to much detail to make good slides. Secondly, tables and illustrations photographed from printed materials are often illegible when displayed as slides. The lettering will usually be too small in relation to the image area and, particularly in typed reports, the image quality may not be adequate for good legibility on the screen unless clean type and a carbon ribbon have been used.

   Ideally, artwork for slides should be prepared specifically for that purpose, but is often possible to prepare artwork with a dual function in mind. Black-and-white artwork for printed publications can generally be used for slides, provided that the information content is sufficiently simple and the standards given in Chapter 5 are observed. The few situations where this will not be possible are discussed in Section 5.2.12.

6.4.2. Slide formats

The most commonly used slide formats are shown in Figure 6.3A. The standard 5 × 5 cm (2 × 2 in) photographic slide uses 35 mm film and pro-

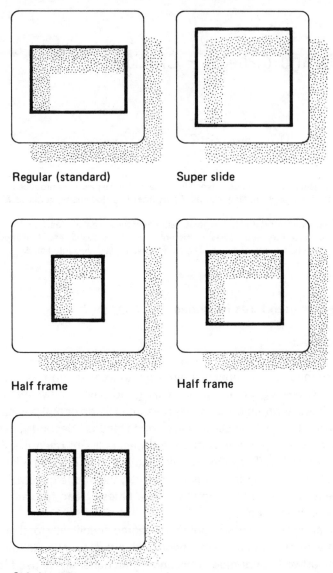

Regular (standard)   Super slide

Half frame   Half frame

Side by side

Figure 6.3A Some common formats for the standard 5 × 5 cm (2 × 2 in) slide (reproduced at approximately 70% of actual size).

jects a rectangular image which is 2 units by 3 units in proportion (see Fig. 6.3B). The image can be in either the portrait or landscape format. An illustration will be able to convey the maximum amount of information if all of this area is utilised. This means that the artwork must have the same 2 × 3 proportion. A square original will waste one-third of the available image area.

Another popular slide format is the 'super-slide'. This also uses a 5 × 5 cm mount, but wider film is used and the projected image is square. The additional image area means that more data can be fitted onto the slide

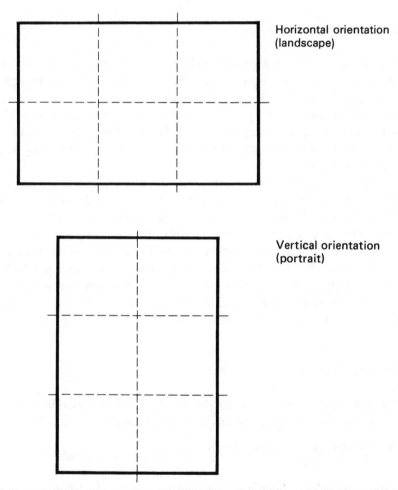

Horizontal orientation (landscape)

Vertical orientation (portrait)

Figure 6.3B Format and shape. A standard 35 mm slide projects a rectangular shape in the proportion of two units by three. This format can be projected with the long axis vertical or horizontal. (In poorly designed lecture theatres only the horizontal format may be accepted.)

without loss of legibility. Super-slides require square artwork, and illustrations prepared for this purpose will often not be suitable for any other media.

The area of a slide can be subdivided if two images are required side by side for comparison. For example, a clinical photograph might be shown on one side and treatment data on the other. Each of these half images is smaller in area than a standard 35 mm image however, and this must be borne in mind in preparing the artwork. Artwork designed for a standard 35 mm slide is likely to be illegible on the screen if it is reduced to fit a half-frame slide.

6.4.3. Size of drafts and artwork

The maximum overall size for artwork for standard 35 mm slides is A4. The reasons for this choice are explained in Section 5.2.3. in relation to artwork for printed publications. If some other size of artwork is used, the 2:3 proportion should be maintained. For super-slides, the artwork should, of course, have a square image area 20 × 20 cm (8 × 8 in). Formulae for calculating the minimum line thickness and minimum capital letter height are given in Section 5.2.4.

If all the artwork for a set of slides is of the same size, this will have the double advantage of speeding production and reducing costs. Preparation of the artwork will be quicker, since it will not be necessary to re-calculate minimum line thicknesses and letter heights for each drawing, and the artist will become used to working within the appropriate standards. When the artwork is photographed, a standard image size in a standard position on artwork of a standard overall size mean that only one camera setting will be necessary. A standard overall size will also facilitate ordering and storage of materials, and storage of finished artwork.

6.4.4. Line thickness

Recommendations for line thicknesses on artwork with a standard 20 × 13 cm image area are given in Section 5.2.4. The formula is:

$$\frac{\text{Longest dimension of artwork (mm)}}{500} = \text{minimum line thickness (mm)}$$

This formula gives a minimum line thickness of 0.4 mm. There is no 0.4 thickness in the micronorm range and in practice a 0.35 mm line weight works equally well.

Too great a variation in line thickness on any one drawing must be avoided for slides. This is because the high contrast photographic techniques often used for black and white originals do not cope well with such variation, nor with large areas of dense black. The suggested micronorm range of line thicknesses between 0.35 mm and 1.00 mm will reproduce satisfactorily, but further differences may not be acceptable (see Section 5.2.12.).

6.4.5. Character height

The minimum capital letter height for legibility at the back of a large lecture theatre can be determined from the following formula:

$$\frac{\text{Longest dimension of image area (mm)}}{40} = \text{minimum cap. letter height (mm)}$$

Using the standard image area of 20 × 13 cm, this would give a capital letter height of 5 mm. The IBM Directory typeface will be adequate, but if a smaller typewriter face is used, then a smaller image area will be necessary. Alternative image areas, line thicknesses and minimum capital letter heights are given in Figure 6.4.

It should be remembered that the recommendations given here are for the *minimum* capital letter height. There is no reason why larger sizes should not be used, and it is often advantageous to use two or three different sizes on the same artwork in order to emphasise important information.

6.4.6. Character style

Methods of lettering suitable for slides include dry-transfer lettering, stencilled lettering and certain kinds of typewriting.

Whatever the method of lettering used, a simple, sans serif style with full, open counters should be chosen. The width of the characters in proportion to their height should be generous, and condensed styles should certainly be avoided. These precautions will ensure that the letters will not fill-in as a result of photographic reproduction and projection. For the same reason, the stroke width should not be too generous, but on the other hand it should not be less than the recommended minimum line thickness. This will ensure that parts of lines will not be lost during reproduction, and that the lettering will be bold enough to be legible when viewed in slide form.

McCormick (1976) has suggested that the width to height ratio of charac-

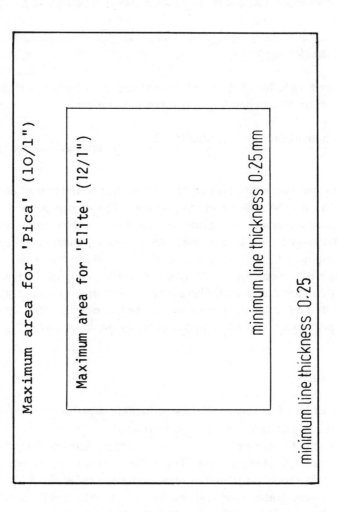

4-5mm cap.ht   minimum line thickness 0.35mm

Maximum area for 'Pica' (10/1")

Maximum area for 'Elite' (12/1")

minimum line thickness 0.25 mm

minimum line thickness 0.25

Maximum area for IBM with 'Directory' type face

ters for use on slides should ideally be 1:1, but that this can be reduced to 1:1.6 'without serious loss in legibility'. The optimum stroke-width to height ratio will depend on whether dark lettering is being used on a light background or vice versa. McCormick suggests a ratio of between 1:6 and 1:8 for black lettering on a white ground, though this is perhaps somewhat generous. A ratio of 1:10, which is achieved with a 5 mm character having a line thickness of 0.5 mm, will be adequate for legibility. McCormick does make the valid point that a slightly reduced line thickness will be preferable for negative images, because of the irradiation and 'flare' which will occur around the edges of a bright image against a dark background. In this case he recommends a ratio of between 1:8 and 1:10.

The width to height and stroke-width to height ratios for a number of common type styles are shown in Figure 6.5, which is given by Galer (1976). Most of the examples fall within McCormick's suggested range of 1:1 and 1:1.6 for width to height ratio. In the case of stroke-width to height ratio, four styles fall within McCormick's recommendations for black characters on a white ground. These are Futura Medium, Helvetica Light, Univers 55 and Univers 53. Characters with greater stroke widths will be in danger of filling-in. For white characters on a black ground, Folio Light and the micronorm stencil system meet McCormick's suggested standards.

The use of capital letters on slides is not recommended, except for the first letter of headings or proper names. The reasons for this were given in Section 1.3.5.

Style	Width: height ratio	Strokewidth: height ratio
Grotesque 216	1:1.6	1:4.5
Helvetica Medium	1:1.4	1:4.8
Futura Medium	1:1.8	1:6.0
Univers 65	1:1.5	1:4.8
Folio Medium	1:1.5	1:5.7
Folio Medium Extended	1:1.3	1:5.3
Helvetica Light	1:1.4	1:7.1
Folio Light	1:1.6	1:9.1
Univers 55	1:1.6	1:7.4
Univers 53	1:1.3	1:6.8
Rotring stencil system	1:1.4	1:10

Figure 6.5 Width to height, and stroke-width to height ratios for a number of type styles (after Galer, 1976).

←

Figure 6.4 Image areas when using different typewriter styles. The format (shape) is the same in all cases and is suitable for standard 35 mm slides. (Image areas shown full size)

6.4.7. Character spacing

The points made in relation to character spacing for printed publications (see Section 5.2.6) also apply here. In the case of slides, however, it is especially important that the lettering should not be too closely spaced, or the letter forms may appear to merge together when displayed on the screen. This will be particularly true for light images displayed on a dark background. Character spacing should never be varied in order to fit an over-long message into a single line, or to expand a short message to fill a line. The spacing should remain constant at all times.

6.4.8. Layout for word slides

The maximum acceptable number of characters per line for slides is approximately forty, or about six words. Allowing for adequate line spacing (see Section 5.2.7), this means that about ten lines of type could be accommodated on a landscape slide, and fifteen on a portrait slide. This would give a total of about sixty legible words on a landscape slide. This does not necessarily mean that such a large quantity of text is desirable however. A more suitable number of lines would be five or six, or about thirty words. Unjustified setting is recommended for slides, in order to avoid hyphenation and variations in character and word spacing.

Headings and text should be ranged left rather than centred. Whatever the method of lettering used, centred lines take more time to produce because their exact length must be calculated before their starting position can be known. They are also less easy to read because of their variable starting position.

Underlining should be used very sparingly. In typewritten lettering it may sometimes be used to give emphasis to a heading or a word in the text, but this is one occasion where capitals may be preferable. When underlining, at least four character stroke-widths should be left between the underline and the character descenders.

Rules are mainly of use in tables, where they should be used to aid horizontal scanning. The use of rules in this way is discussed in Section 3.2.2.

6.4.9. Tones and shading

Minimum dot size and spacing between cross-hatched lines can be calculated from the formulae given in Section 5.2.4.

Many of the comments made in Section 5.2 in relation to printed materials also apply here. It must be remembered, however, that large areas of dense black may not reproduce well using high contrast photography.

6.4.10. Positive versus negative images

Black characters on an intensely white background may cause discomfort as a result of glare. Bright white characters on a dense black background will suffer from irradiation and 'flare' however, and may not be so legible unless special precautions are taken in relation to the design of the characters and their spacing.

This question of image polarity has been investigated by Morton (1968). He prepared a series of five different slide forms, which consisted of black lettering on a white ground, black on yellow, white on black, yellow on black and white on blue. The letter forms used were those specified by British Standard 4189 for the testing of vision. Morton identified the smallest character size at which subjects made no recognition errors. His results are shown in Figure 6.6. He found that for subjects with the poorest visual acuity, positives were the most legible. As a general rule it would therefore seem advisable to avoid negative presentations. In some cases, however, a negative image may be an effective way of drawing attention to small areas containing captions.

Whichever image polarity is chosen, the slides in any one presentation should be consistent. Constant change from light to dark backgrounds will prevent the eyes from becoming adapted to any particular level of lighting and will cause fatigue. If colour is used for image or background, a high contrast must be maintained if legibility is not to be impaired.

	Slide form		
Legibility			Subjective preference
(best)	Positive, black on white	Positive, black on white	(38%)
	Positive, black on yellow	Diazo, white on blue	(27%)
	Diazo, white on blue	Positive, black on yellow	(26%)
	Negative, white on black	Negative, white on black	(5%)
(worst)	Negative, yellow on black	Negative, yellow on black	(4%)
			(100%)

Figure 6.6   Ranking for legibility and subjective preference, for five slide forms (after Morton, 1968).

99

6.4.11. Contrast

The level of contrast on a slide is a very important factor in its legibility. Wilkinson (1976) found a significant decrease in legibility when the contrast ratio for dark figures on a light ground was reduced from 1:95 to 1:60, and from 1:60 to 1:30. He also found that there was a strong interaction between contrast and letter height, changes in either having the greatest effect when the other was at its minimum.

It is therefore essential that all artwork should be prepared with suitable materials and by suitable photographic methods in order to ensure maximum contrast.

6.5. The use of colour

6.5.1. Black-and-white versus coloured artwork

Black-and-white artwork can be used to produce black-and-white slides or black and one colour, depending on the photographic process used. If diazo film is used, black-and-white artwork can be used to produce a white image on a coloured background (often blue). Coloured artwork can be used to produce full colour slides if colour film stock is used.

Black-and-white artwork has the advantage of being more flexible than colour artwork. Corrections, alterations and additions can be made easily by 'cut-and-stick' methods on black-and-white artwork. So long as a high contrast photographic method is used, the alterations will not be visible on the projected image.

Coloured artwork photographed on colour stock requires great care in preparation however, as any flaw will appear on the slide. Cut-and-stick corrections, erasures and dirt will all be visible. Cleanliness of coloured artwork is vital.

6.5.2. The use of coloured images

The use of colour is a realistic possibility on slides, whereas it is usually too costly for use in printed publications. Colour will certainly increase the attention-attracting qualities of a slide, and it can be used with good effect to emphasise, divide and relate information.

When using colour on slides, it is important to choose highly saturated or 'pure' colours as these are easier to discriminate. Relatively unsaturated mixtures such as pinks, khakis and browns tend not to work well on slides, and metallic colours such as gold, silver, copper and bronze should be avoided too. It must also be remembered that colour saturation on slides can be seriously diminished by inadequate power from the projector lamp, and by stray light falling on the screen. The best colours to use, therefore, are highly saturated reds, oranges, yellow, greens and blues.

Colour on slides can be used for lettering, lines or shading, but the effects of the resulting image and background colour combinations on the legibility and discriminability of the coloured images must be borne in mind. In terms of the legibility of lettering, the most important factor is the brightness contrast, or tonal difference, between the image and the background. For optimum legibility, this contrast should be as high as possible. This would imply that 'dark' colours having a low reflectance should be used on 'light' backgrounds, and 'light' colours having a high reflectance should be used on 'dark' backgrounds. Mid-tone backgrounds such as grey must be selected carefully. Where the image and background colours are similar in tone (or reflectance), optical interference effects may be created. Red lettering on grey of a similar reflectance is particularly difficult to read.

It is important to realise, however, that colour discrimination will be affected by the level of tonal contrast between the image and the background. This will be particularly important where several different colours are used to code information, as in the case of coloured graph lines. Coloured lines on a white background will appear darker in tone than the same lines on a black background. Thus a 'pale', relatively unsaturated yellow will appear 'darker' on a white background than the same colour on a black background. In general, however, colour is destroyed by the strong tonal contrast which will exist on a white background. 'Dark' backgrounds (not necessarily black) are therefore preferable where colour discrimination is important.

The size of the coloured areas is also important, because the human eye is not capable of discriminating between small areas of colour. As the area diminishes, so colour effects are lost and only tone effects remain. This will be especially true for graph lines on a white background, where colour discrimination will also be impaired by the contrast effect mentioned above. Thus a thin red line and a thin blue line on a white ground may both appear as 'dark' or even as black lines which are indistinguishable in terms of hue. This difficulty can be counteracted to some extent by increasing the line thickness and the stroke thickness of coloured lettering on white backgrounds.

If colour is used as a coding system, the number of different colours should be kept to a minimum to ensure that they can be easily discriminated from one another. Four is probably a realistic number of colours for use on slides, and the colours chosen should be as different as possible in both hue and tone. Blue and green, for example, might easily be confused because they are relatively close together in terms of wavelength, or hue. The possibility of confusion could be reduced by choosing a blue towards the violet end of the spectrum, and a yellowish green, thus increasing the difference in hue. Difference in tone will facilitate discrimination still further, particularly for those with colour vision defects. Approximately 8% of men and 0.4% of women have some form of anomalous colour vision, the most common problem being the discrimination of greens from reds. Another way of ensuring discrimination between differently coloured areas is to use mechanical tones such as stippling or cross-hatching. The colour is then redundant, but serves to reinforce the distinction between the areas.

6.5.3. The use of coloured backgrounds with black lettering

The use of coloured backgrounds with black lettering has been investigated by Snowberg (1973). He used two sets of red, blue, green, yellow and white backgrounds. One set was matched for luminosity and the other for transmission density. Each slide contained five lines of randomised black letters. Five different letter heights were used, giving capital letter height to screen height ratios of 1:20, 1:30, 1:40, 1:50 and 1:60.

Snowberg found that when the slides were matched for luminance there were significant differences between the hues in terms of legibility. The effects were particularly marked for the smaller letter sizes of 1:50 and 1:60. A white background was significantly better than yellow, green, red and blue, next came a yellow background which was significantly better than green, red and blue, and blue was significantly worse than the other four backgrounds. Researchers in optometry have shown that green, and especially yellow, focus on the fovea of the eye. Red focuses beyond the fovea and blue focuses in front of it. These studies of chromatic aberration explain the fact that even when luminances are equal, there are still differences in acuity between the colours. No such clear cut hierarchy was detected on the slides matched for transmission density, though there were significant differences between the colours with the smaller letter sizes. Snowberg concluded that the hierarchy depends on the interaction of colour and brightness, or luminance. Figure 6.7 shows the rank order for acuity of the ten backgrounds tested, and their brightness (differences between adjacent ranks 2 to 9 were not significant).

Snowberg therefore recommended white backgrounds for maximum visual acuity, and suggested that blue backgrounds should be avoided where legibility is critical.

Rank	Colour	Brightness level
1	White (L)	15
2	Yellow (L)	15
3	White (T)	10
4	Green (T)	16
5	Blue (T)	19
6	Red (L)	15
7	Yellow (T)	10
8	Green (L)	15
9	Red (T)	10
10	Blue (L)	15

Figure 6.7 Slide luminance values arranged hierarchically from highest to lowest overall acuity (after Snowberg, 1973). (L = equal luminance; T = equal transmission density.)

6.6. Testing slides

The only satisfactory way of checking slides, whether in black and white or colour, whether for legibility or colour saturation, is to view them from the back of a large lecture hall. To check slides at close quarters in a small room and on a small screen will not be adequate, because the projected image will be much brighter than it would be when projected from a greater distance onto a larger screen. Colour slides which appear to be very distinct at close quarters can be very disappointing when viewed in a large lecture hall. Similarly, lettering on slides which is clearly visible in the office or seminar room may be totally illegible when viewed in the lecture theatre.

6.7. Projection of slides

The quality of the projected image will depend on such factors as the size of the screen image in relation to the distance of the farthest seating, the angle at which the screen is viewed and the level of illumination in the auditorium during projection.

Image size is determined by the distance from the projector to the screen and the focal distance of the objective in the projector. Most manufacturers supply tables which relate the projector/screen distance to the image size. It is important to distinguish between the size of the projected image and the size of the screen. The image size must be such that a standard 35 mm slide is legible from the farthest row of seats. The size can be described in terms of the longest dimension of the image area, since it does not matter whether the format is vertical or horizontal as far as legibility of the data is concerned.

For optimum legibility, the first row of seats should not be closer than a distance of twice the longest dimension of the projected image, and the last row should not be more than six times this length away from the screen (see Figure 6.8). The relation between viewing distance and the minimum size of projected lettering necessary for legibility is shown in Figure 6.9.

The size and shape of the screen required will depend on the slide formats to be accommodated. Although the majority of users seem to prefer horizontally oriented slides, there are many occasions when a vertical format is used and the screen must therefore be large enough for either format. The screen should also be wide enough to accommodate the projection of two super-slides (square format 35 mm slides) side by side, as shown in Figure 6.10. The image should adequately fill the screen area available if best use is to be made of the space in the auditorium. The height of a vertical image should be approximately equal to the height of the screen. If the image is too large or too small, as shown in Figure 6.11, then the projector should be moved or the lens changed.

Viewing angle has a very important effect on legibility. Wilkinson (1976) carried out an experiment in which he tested nine different viewing angles, at 15° intervals on an arc from 60° to the right and left of the subject. He found a significant detrimental effect on legibility as the viewing angle deviated from 0°. Ideally, all spectators should be seated within an arc of 30° to the left and right of the screen, as illustrated in Figure 6.10.

Partial illumination of the auditorium is often desirable so that the audience can make notes, but excessive light will reduce contrast and desaturate colours in the projected image, thus impairing legibility.

6.8. The design of lecture theatres

Teaching staff are often asked to advise on the design of a lecture theatre or the purchase of equipment for it. Medical illustrators, projectionists and audio-visual technicians are at worst not consulted at all, or at best consulted after the theatre has been built and the equipment purchased! It is therefore important that teaching staff should be aware of some of the basic problems. Questions such as the position of the projection equipment and whether or not air conditioning should be used are not dealt with here, since these and similar problems tend to be specific to individual situations. The following points, however, should be considered in the design of any lecture theatre:

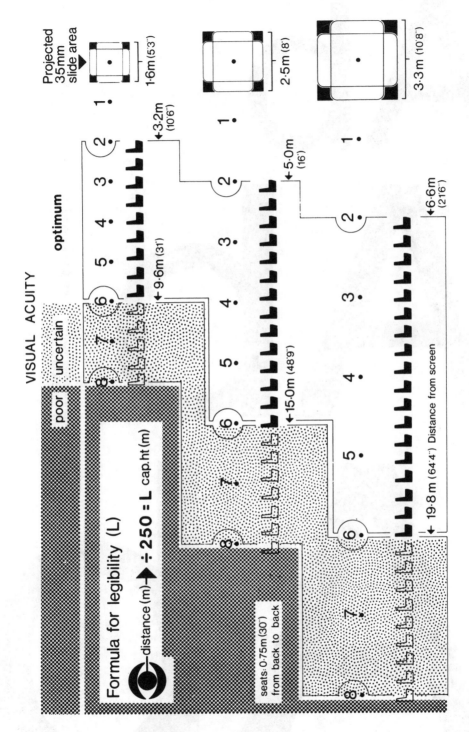

Figure 6.8 Visual acuity and the viewing distance.

4 cm Ab

6 cm Ab

8 cm Ab

Figure 6.9. Minimum acceptable size for lettering as projected onto the three screen sizes given in Figure 6.8 (reproduced actual size).

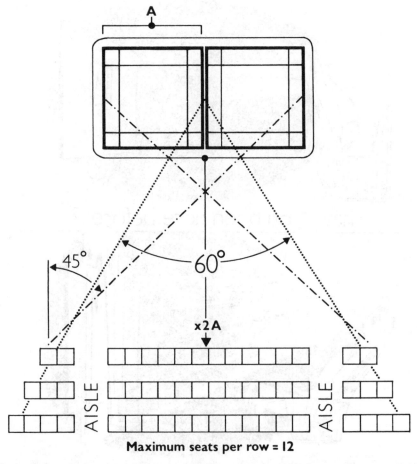

Figure 6.10 An example of a screen capable of accepting any slide format up to two side-by-side super-slides. Note that viewers on the extreme sides of the auditorium may receive considerably less illumination from the screen and thus find data at the edges of the slide difficult to read if the lettering is not large enough.

1. Acoustics should be considered in the early stages of the design. If the architecture is not conducive to good acoustics, then it is highly unlikely that the subsequent addition of padding or baffles will improve the situation. It must be remembered that *some* reflected sound is necessary or the speaker would be unable to hear questions from the audience.
2. A lecture theatre which is restricted in size to 70 cubic feet per seat with a maximum of 300 seats is ideal for the unaided human voice.
3. A rostrum will improve acoustics by raising the speaker, but it may interfere with other audio-visual requirements.

have you been here before⌐?

Figure 6.11 Two cardinal errors in the design of lecture theatres. Either the projector lens is unsuitable or the projector is the wrong distance from the screen. (It should always be possible to project slides with the long axis vertical as well as with the long axis horizontal.)

4. Padded seats and cork floors help acoustics.
5. Background noise levels from air conditioning systems, projectors, etc. should not exceed 10 decibels when measured in an empty theatre.
6. It is a mistake to place entry/exit doors in the same wall as the rostrum, screens, blackboard, etc. as this will reduce the amount of space available for audio-visual equipment.
7. Ventilation systems should not occupy valuable space on the 'teaching' wall of the theatre.
8. A central aisle should be avoided, as this is the best position for viewing the projection screen.

9. The maximum number of seats per row between aisles should not be more than twelve.
10. The raking of seats need only be sufficient to give a clear line of vision above the heads of those sitting in the next row forward. A difference in height of about 20 cm (8 in) will be adequate. Raked seats may not be necessary for auditoriums seating less than 100 people.
11. A foyer or anteroom through which the audience has to pass before entering the theatre provides an important 'adjustment' space. It will help allow the eyes to begin to adapt to the darker environment within the theatre, and it will also encourage the audience to enter quietly.
12. The most important and often the most neglected factors are the distance between the projector and the screen, and the distance between the audience and the screen. The interrelation of these factors was discussed in Section 6.7. above.

7. Posters

7.1. Introduction

The use of posters as a means of disseminating research results is becoming increasingly popular. International conferences, for example, often organise poster sessions which delegates can visit in breaks between the more formal sessions. A number of research workers will generally be invited to take part in a poster session. Each will be expected to prepare a poster or a series of posters summarising his work, and to be present at the session to discuss the subject with interested colleagues. Restrictions on display space – and of time on the part of the visitors to the session – usually mean that only a very limited amount of detail can be given, and the presence of the presenter is normally essential.

A poster display functions well if it attracts the attention of passers by, and if the author is able to withstand the ensuing discussions which can often be controversial and animated. A poster display does not function well if it is 'unattractive' and therefore not looked at closely, or if the presenter is shy of argument and discussion or not present.

Some researchers have expressed their dislike of presenting a paper in a poster session for the very reason that they are not protected from audience participation to the extent that they would be in a formal lecture. In spite of the fact that formal lectures usually have a question period, the speaker is not vulnerable for very long. In a poster session, however, he or she is vulnerable throughout. Nevertheless, many researchers agree that there is much to be gained from face to face meetings and discussions with persons interested in the same subject. The presenter in particular is likely to gain much more from this direct contact than he or she would from a more formal situation where the communication is essentially one way.

Some organisers may group the more 'important' posters in one room where semi-formal talks are also part of the poster presentations. In other rooms the less important posters will be displayed on a more casual basis. In this latter case, visitors are able to go to their own area of special interest without having to wait for a specific time as is often the case in formal meetings.

The disadvantages of the poster session are several. The limited amount of space available and the fact that the lettering needs to be large enough to be readable from about 1 m (3 ft) mean that the quantity of information which can be presented is somewhat restricted. A further disadvantage is that certain positions in the room will be much more favourable than others. Positions near doors are likely to 'capture' an audience, and positions near windows are often better lit and make the display easier to read. Darker areas in the corners of the room tend to be disadvantageous.

7.2. The content of posters

7.2.1. General principles

In studying a poster the reader can linger over it as long as he or she wishes, and in this sense a poster presentation is more like a printed page than a slide. In most cases, however, the reader will be confronted with a large number of posters, and he will be expected to study them while standing up. Information must therefore be presented clearly and concisely so that the essentials of the message are easily grasped. Studies in museums have shown that visitors tire of reading long caption panels very quickly, and while they may begin by systematically reading each one, they soon start to wander at random. After this point they are likely to spend more time on the most visually 'attractive' displays. Posters should stimulate interest rather than present complex details.

In effect, a poster is an advertisement for the author's particular ideas or techniques. He stands in front of his display with something to sell. Good posters use the best techniques of salesmanship. The content should be succinct and to the point, with short pithy subheadings, and the design should be attractive in terms of colour, lettering and layout.

7.2.2. The structure of the poster

As with slide presentations, it is usual to divide the information content of the poster into at least three basic units or panels, i.e. an introduction, the body of the presentation and a conclusion.

The observer is likely to look first at the poster heading, and then at the conclusions. Only after having satisfied himself that the material is interesting will he bother to read the rest. The introduction and conclusions therefore need to be short and to the point, and presented in an attractive and easy to read style, as in Figure 7.1.

Introduction

CONGENITAL PARATHYROIDISM is common but not easily recognised.

NEW TECHNIQUES of investigation are demonstrated which change the patterns of diagnosis.

Conclusions

- Serial blood tests following injection with 'Isopon A' prove a useful and safe indicator.

    - Ultrasound and isotope scanning provide 99% accuracy in diagnosis.

        - Therapy with 'Dilactyl' AT THE RIGHT TIME, is shown to be completely effective in all fifty cases investigated.

Figure 7.1   The importance of simplicity when presenting introduction and conclusion data on posters.

7.2.3. Headings

Poster subject headings, like slide headings, need to be as short yet as meaningful as possible. Presenters are sometimes asked to give their posters the same title as the abstracts which will subsequently appear in the 'Proceedings'. This can result in impossibly long titles which cannot be displayed in an eye-catching manner and which therefore defeat the main purpose of the title card. Thus a heading such as 'An in-depth study of fifty cases of postoperative congential parathyroidism' would be much better abbreviated to 'Congenital parathyroidism – 50 cases'. The short heading is more likely to attract attention and its meaning is easier to grasp.

The choice of words for subheadings is important too. 'Introduction' and 'Conclusions' are automatically sought after, but other headings can either repel or attract interest. Short and simple words are best. Some words are universally evocative and others are not. For example, a subheading to a section describing 'cell lysis in haemolytic disease' could be simply headed 'Cell destruction' or even more simply 'Destruction'. The technique of choosing evocative words is demonstrated in every daily newspaper; they are deliberately chosen to capture interest and compel reading.

7.3. Useful equipment for poster production

Many people presenting data at poster sessions have to prepare their own posters, and often at the last minute. In a well-equipped illustration studio it would be possible for amateurs to produce a good-looking poster using colour and quality lettering within about 4 to 6 hours, though it might require the efforts of two or three people. A technique for doing this will be described for a poster filling an area 1.80 m × 1.20 m (6 ft × 4 ft) (see Section 7.4.2.).

To prepare a quality poster in less than a day, the following equipment would be desirable in addition to normal drawing office materials:

1. An Econosign tape lettering machine.
2. An IBM Executive typewriter with 'Directory' typeface.
3. A reflex printing machine.
4. A photographic means of obtaining enlarged prints. A process camera and the Agfa Gevaert 'Copyproof' system will produce instant results, and will avoid the delays often associated with obtaining prints by normal photographic methods.

7.4. Graphic standards for posters

7.4.1. Poster formats

Ideally, the overall size of posters should be determined by the quantity of information which the reader can be expected to absorb, and the size of lettering which will be necessary for legibility at the average distance from which the poster will be read. In practice, however, the size of posters is generally predetermined by the organisers of the meeting and is dependent on the number of exhibitors in relation to the total display area available. Instructions from organisers of poster sessions are often inadequate. For example, the overall dimensions of the display area alotted to a particular presenter may be specified, but the orientation of the display rectangle may not be given. It is therefore important for the presenter to find out whether the display rectangle has its long axis horizontal (landscape) or vertical (portrait).

When the presenter has some freedom of choice, the use of one of the International Standard Paper Sizes will have two advantages. Firstly, because the different sizes have similar proportions, posters made up of different sized units can be successfully displayed together if necessary. Secondly, these proportions are convenient for other forms of reproduction.

7.4.2. Size of drafts and artwork

It is generally easier for amateurs to plan poster layouts on a 'same-size' basis, i.e. when the artwork is the same size as the finished poster. Using the equipment recommended in Section 7.3., this is a realistic possibility. If, however, headings are conventionally typeset then same-size setting will be expensive, and if the text is IBM typeset, enlargement will be essential. In these cases, therefore, it would be necessary to prepare the artwork on a smaller scale and then enlarge it.

Assuming that same-size artwork is a possibility, then the artwork for a 1.80 × 1.20 m (6 × 4 ft) poster in landscape format might be planned in the following way. Such a large display must first of all be broken down into modular units for ease of handling and transportation. The number of units required will depend on the overall size of the display, and their exact measurements will be influenced by the sizes of card or paper available. A suitable size of card for a 1.80 × 1.20 m display is 110 × 80 cm (43 × 31 in). This can be divided into four units which, when trimmed, will measure approximately 38 × 54 cm ($15\frac{1}{2}$ × $21\frac{1}{2}$ in). This is a convenient size for carrying. Eight such units mounted in portrait format in two rows of four will occupy most of the 1.80 × 1.20 m (6 ft × 4 ft) display area, while still allowing a 13 cm (5 in) horizontal strip at the top of the display area for the title card (Figure 7.2).

Many poster session organisers make the mistake of encouraging presenters to create such large title cards that a quarter or more of the available display space is taken up with the subject title, the author's name and his institution. This is unnecessary (see Section 7.4.3 below). No title card need be wider than 13 cm (5 in), though it may be as long as the display space will allow. Using 110 × 80 cm card, the maximum length would be 110 cm. The card can be scored on the back and folded face to face; this reduces its length to 54 cm (21½ in) for packing (Figure 7.3).

The 38 × 54 cm (15½ × 21½) units form the mounts for the data. Two A4 (8 × 11 in) sheets of text or illustrations and two 15 × 10 cm (6 × 4 in) prints,

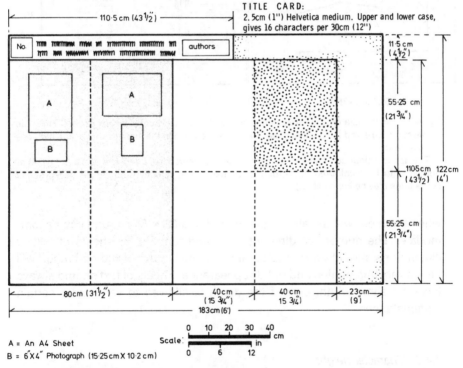

A = An A4 Sheet
B = 6˝ × 4˝ Photograph (15·25 cm × 10·2 cm)

Figure 7.2 Planning poster sessions. In this example, the display area has been taken as 183 cm (6 ft) by 122 cm (4 ft). The solid area represents one standard-size (UK) sheet of card. The stippled area represents one of the eight pieces which make up the display. The whole display is now reduced to a modular system which is easy to carry under the arm or pack in a suitcase. The title card is made from an 11 cm (4½ in) wide strip cut from the long edge of a full-size card. This is scored on the back and folded to make it fit the final pack.

Score the BACK of the title card lightly with a knife, then fold face to face

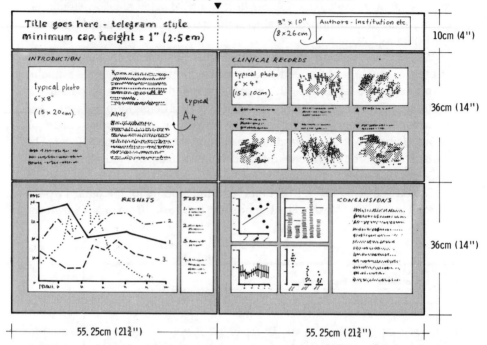

Figure 7.3 In this example the area allotted for the display is 112 cm (44 in) by 82 cm (32 in). The whole standard-sized (UK) sheet of card is now subdivided in the manner shown and includes the title card.

The use of colour can make a poster session more attractive. Longer texts can be typed or printed onto light coloured backgrounds. (When typing directly onto colour backgrounds corrections may be impossible.)

when trimmed, will usually be the most that a 38 × 54 cm unit can accommodate. The number of different sizes and formats for sheets of text or illustrations should be kept to an absolute minimum, or the end result will have a very fragmented and untidy appearance. Sheets of text should always have the same width, even if the depth is allowed to vary under certain circumstances.

7.4.3. Character height

Posters will generally be read at a distance of about 1 m, as opposed to the normal reading distance of about 35 cm (14 in) for books. The character size will therefore need to be correspondingly larger for posters. Typescript from a standard typewriter, computer printout or figures taken directly from most

116

printed materials will not be adequate.

In the presentation of data, all general texts and symbols on a poster should be legible from the same distance. The reader should not have to move closer to peer at diagram captions. Some organisers of poster sessions ask for a size of lettering which will be legible at a distance of 2 m ($6\frac{1}{2}$ ft). In most cases, however, visitors will be discussing the poster with the presenter at a distance of about 1 m, and lettering of the size demanded will therefore be unnecessarily large. A suitable capital letter height will be between 6 and 8 mm. Even 4 mm, though not ideal, can be used provided that the lines are adequately spaced.

The title card needs to be legible from a greater distance, in order to attract visitors searching for posters on particular topics. In this case the lettering should be legible from about 5 m (16 ft) or more. This will require a minimum capital letter height of 4 cm ($1\frac{3}{4}$ in). It is only the title of the presentation which needs to be displayed in this size however. Although the author's name and institution are an important part of the title card, they do not necessarily need to be in lettering any larger than the text size. The use of larger lettering may result in an unnecessarily deep title card, thus reducing the area available for the presentation of the data. The data should take precedence over personal advertising.

It is also desirable that subheadings should be legible from a greater distance than the text, as these also tend to attract interest. A capital letter height of 10 to 16 mm (3/8 to 5/8 in) will be suitable. Capitals should be avoided (see Section 1.3.5).

7.4.4. Character style

Most of the comments made in Section 6.4.6. in relation to slides will also apply here. A simple sans serif face will generally be preferable. It should not be a condensed face, and the weight should tend towards bold rather than light. The final choice of character style will, of course, depend on the method of lettering used. The alternatives for the text will be IBM Directory typescript, stencilled or dry-transfer lettering, or enlarged typesetting. The latter option will obviously give the greatest choice of typeface, but will be more costly and may result in delays if it cannot be done in-house.

For title cards and subheadings, Econosign, dry-transfer or stencil lettering will be the most suitable. If IBM Directory is used for the text, appropriate dry-transfer lettering typefaces would be Univers or Helvetica.

Both text and headings should be in upper and lower case lettering. The extensive use of capitals should be avoided for the reasons given in Section 1.3.5.

7.4.5. Layout for text

Many of the comments made in relation to slides in Section 6.4.8. also apply here.

If Directory typescript is used for the text, a line length of 15 to 18 cm (6 to 7 in) will be suitable. If this is typed onto paper which is 28 cm (11 in) wide, this will give adequate margins for trimming. If some other form of lettering is used, lines of about forty characters in length will be appropriate.

7.4.6. Illustrations

The importance of logic and simplicity in the design of tables and figures has already been stressed. There are, however, some additional points worth noting in relation to posters.

Whenever possible, captions on graphs and diagrams should be horizontal. Vertical captions will not seriously inconvenience a seated audience looking at slides, but in the case of posters the audience will be in a standing position and will be more easily tired. They should therefore be spared the inconvenience of having to twist their necks in order to read captions.

Artwork prepared for other purposes (such as slides or in some cases printed publications) will often be suitable for use on posters if it has been prepared in accordance with the standards suggested in this book. Drawings in an A4 format ($8\frac{1}{4} \times 11\frac{3}{4}$ in) may not need further enlargement. Similarly, photographic prints which are 15×10 cm (6×4 in) and which show details clearly at normal desk top viewing distance may not need enlarging. It is important, however, that prints should be perfectly flat and not show any bumps or have buckled edges or surfaces, otherwise subsequent mounting can be difficult. All illustrations should be trimmed to present only the essentials.

Original illustrations should not be used for posters as they can be irreparably damaged when the display is dismantled. Black-and-white originals can be copied on a reflex printing machine, which gives a same size facsimile print of the original. If colour is required, this can be achieved with felt-tipped markers or coloured transparent overlays.

7.4.7. Poster layout

The division of information between the component units of the display should be related to its structure as far as possible. Ideally each unit should

118

have its own subheading, though in some cases more than one unit will be required under a particular subheading. If there is relatively little information under a particular subheading, it will often be better to leave the rest of that unit blank, rather than to split subsequent subsections unnecessarily between units. The logical sequence of the units themselves should be immediately obvious. In a display consisting of two rows of four units each, spacing between the units can be used to indicate to the observer whether he should read across the rows or down the columns. The use of numbered subheadings will also help. Within each unit, related text and illustrations should be linked spatially, and great care should be taken where tables or illustrations cut across more than one column of text (see Section 1.10).

If IBM Directory typescript is used for the text, the simplest way to plan the layout is to use typed drafts of the text. These drafts can then be arranged on pieces of paper or cards which are the same size as the final units, or a small sketch plan can be made showing the relative position of each item on its particular card. It will be important to identify the card units with a letter and the items to be fixed on them with a serial number. When the layout has been arranged, these identifying marks can then be written on the back of each finished piece of artwork, text or photograph. They must be written clearly in the *centre* of the item so that they are not inadvertently removed during trimming processes.

Each card unit should be carefully checked, and the text and illustrations then finally mounted using spray glue adhesives (Section 11.2.10).

7.4.8. The use of colour

The use of colour on posters can be valuable for attracting attention. An experiment by Dooley and Harkins (1970) showed that more attention was paid to coloured posters, regardless of whether the colour was used meaningfully or not. If colour is used, however, we strongly recommend that it should be used logically and with discretion.

In poster displays made up of card units as described above, the mounting card can be coloured. A strong or dark colour will be suitable. Texts and illustrations can then be prepared on light coloured paper. Dark papers should not be used for the information itself, as they will reduce the contrast between the lettering and the background and thus impair legibility. Different coloured papers may be used for different units. For example, the text in the introduction and conclusions units might be prepared on paper of a brighter colour from that used in the rest of the display. Alternatively, subheadings might be typed on a colour different from that used for the body of the text. It should be remembered that if a dark coloured mounting card is used, text or headings on white or yellow paper will stand out more than any other colour and will seem to be the most important items.

If the text cannot be typed directly onto coloured paper (for example if it needs enlargement), or if reflex prints are used, colour can be introduced by means of felt-tipped markers or transparent adhesive overlays, as suggested above. Direct enlargements of lettering and diagrams in colour are possible using Agfa Gevaert copychrome materials.

7.5. Packing and transporting poster displays

The eight-unit poster display described above is designed to be easily packed and carried flat. Posters should never be rolled, as this creates problems when they are next required for use. Artwork, photographs and lettering may stick to adjacent surfaces in the roll, and dry-transfer lettering is particularly liable to damage. Variations in temperature and humidity such as those which occur in aircraft holds will greatly increase the chances of damage to dry-transfer lettering. A flat 38 × 54 cm package can always be taken as cabin baggage, and is less likely to come to harm.

8. Overhead projection transparencies

8.1. Introduction

There is a considerable amount of literature available on the techniques of overhead projection. What follows here is a summary of some of the basic requirements and limitations.

OHP is an extension of the chalk board, and as such is ideal for use in small seminar groups. The equipment and screens are not generally suitable for use with large audiences however. Few screens are larger than 6 × 6 ft, and they usually have to be angled down towards the projector if distortion of the image is to be kept to a minimum. For these reasons the maximum audience size will be much the same as that for a chalk board.

OHP has the following advantages:

1. The lecturer faces the audience.
2. Colour can be used at similar cost to black-and-white.
3. Illustrations can be prepared in advance.
4. Transparencies can be re-used many times.
5. Information can be built up piece by piece using the overlay technique, or dismantled in a similar fashion.
6. Certain forms of animated (moving) diagram are possible.
7. Informal and spontaneous 'instant' illustrations are possible, just as with a chalk board.
8. OHP is cleaner than a chalk board and does not create dust.
9. OHP can be used without blacking out the lecture room.

Although OHP is convenient for 'last minute' illustrations, used in the manner of a chalk board, this is a rather limited use of a versatile and effective medium. For lectures which are often repeated to average sized audiences, OHP techniques have special advantages. See especially items 3, 5 and 9 above.

8.2. Graphic standards for transparencies

8.2.1. Originals

There are several techniques which can be used to enhance the effectiveness of OHP transparencies, and these are described in Section 8.3 below. If the use of complex techniques is envisaged, however, it is essential that these should be taken into account at the planning stage.

It is possible to draw directly onto OHP transparancies with specially designed felt-tipped pens. Pens are supplied with broad, medium or fine tips and deliver coloured inks which are either spirit based or water based. Both sorts of ink can be used to advantage on the same drawing. 'Permanent' lines can be drawn in spirit based ink and 'temporary' lines in water based ink applied later, perhaps while the lecturer is talking. These temporary lines may be subsequently removed with a damp tissue without eliminating the permanent parts. Existing illustrations can be transferred to OHP cells by xerox or other processes in black-and-white or colour, though colour copying is usually limited to flat colour line techniques using some form of screen to produce half-tones.

Freehand lettering has to be done carefully if it is to remain legible. Stencil guides and dry-transfer lettering are available. The latter is produced in special 'projectable' quality colours.

8.2.2. Format and size

The format and size of most OHP acetate sheets (or cells) is a square measuring 25 × 25 cm (10 × 10 in).

8.2.3. Line thickness

The standards given in Chapter 6 in relation to standard 35 mm slides will also work well for OHP transparencies, provided that the farthest seats are not more than 5 m (16 ft) from the screen. The screen image must be about 6 × 6 ft in size however. If the image is smaller than this, the size of the lettering must be correspondingly increased if it is to be legible at a distance of 5 m. In this case, the amount of information which can be fitted onto the sheet or cell will be reduced.

8.2.4. Character height

A character height of 5 mm on the cell may give an approximate projected letter height of about 25 mm (1 in) on the average size screen. Projectors, screen sizes and audience distances vary considerably however, so a check from the back row of the seminar room is strongly advised.

8.2.5. Layout

Given that lettering should ideally have a capital letter height of not less than 5 mm and that it needs to be fairly open in character, the number of words which can be accommodated on one sheet of acetate is limited. Approximately forty words per sheet is ideal, but when long scientific words are used, perhaps in conjunction with numerals, it may be better to assess the number of characters rather than the number of words. In this case the design should not contain more than 240 character units (including spaces between words). In relation to the layout of tables, Chapter 3 should be consulted.

8.2.6. The use of colour

Colour can be used very effectively on OHP transparancies, but it should be used functionally rather than decoratively. If, for example, blood vessels are being dealt with, blue should be reserved for veins and red for arteries. With text too, colour should be used logically. Thus certain colours might be reserved for headings, or used to indicate particular kinds of information. Unnecessary colours should not be introduced for decoration, just because they happen to be available.

Coloured overlays or background cells need to be highly saturated. Colours will be desaturated by light from the projector lamp, and also by the illumination within the lecture theatre. Colours which appear garish on the cell may be relatively mild when projected.

8.3. Techniques in the use of OHP transparancies

8.3.1. Overlays

A convenient technique for the use of four overlays is shown in Figure 8.1.

123

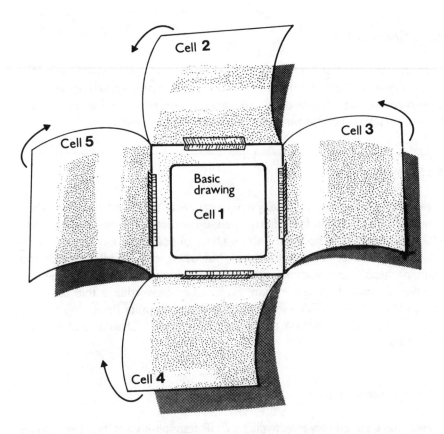

Figure 8.1 Successive overlays of parts of a drawing or text can be used to add data to a basic illustration. This is a very flexible 'additive' or 'subtractive' technique in overhead projection, as the four available cells can be used in any order. Each cell is fixed to the basic drawing by means of a strip of drafting tape. The basic drawing should be simple and bold so that it remains clearly visible even when it is covered by four cells.

Each overlay sheet should be fixed to the background sheet with a long piece of tape. This will ensure good 'register' when the overlays are flapped down over the background. Pin holes are often the best form of registration mark for loose overlays, as they are relatively inconspicuous on the screen.

When overlay techniques are used, the basic drawing must be bold and simple if it is to remain clearly visible through overlays which may themselves carry data.

8.3.2. Reveals

An alternative method of building up data on the OHP screen is to reveal only part of the artwork at a time, as shown in Figure 8.2. The conclusion of a talk can be effectively presented point by point in this manner. The reading speed of the audience is then paced by the lecturer so that due time is given to the discussion of each point. By the end of the conclusion, the full list of points will be visible. This technique can also be used for diagrams, but masking, as described below, may be more suitable.

Figure 8.2   A technique for revealing data in stages for overhead projection. By this means the lecturer can concentrate on each point in turn.
   The data must be in the form of a negative, i.e. as clear lettering or diagram on an opaque background.

8.3.3. Masking

In this technique, masks are made to cover specific parts of the artwork. Build-up is achieved as the masked areas are removed. It is therefore possible for the lecturer to discuss each part of an illustration independently of the remainder, and he can discuss the various parts in any order he wishes.

   Masking techniques are most effective if negative images are used, i.e. white or coloured illustrations on a black background. The masking sheets will then be of clear foil, with opaque sections where the masking is required. In the example shown in Figure 8.3, when all the sheets are folded over the

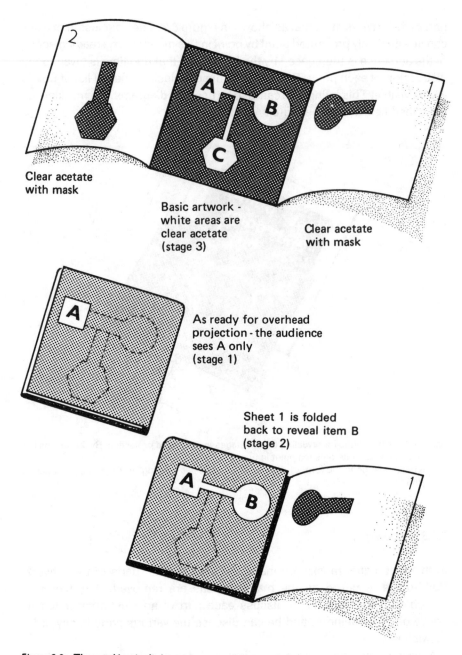

Clear acetate
with mask

Basic artwork -
white areas are
clear acetate
(stage 3)

Clear acetate
with mask

As ready for overhead
projection - the audience
sees A only
(stage 1)

Sheet 1 is folded
back to reveal item B
(stage 2)

Figure 8.3 The masking technique.

126

artwork, only square A will be visible. Removal of sheet 1 will reveal circle B, but sheet 2 still masks rectangle C until it is folded back.

Negative image illustrations and masking techniques have many variations which an imaginative user can develop. For example, graph lines can be animated simply by pulling a mask to one side as shown in Figure 8.4. A variant of this technique is to apply low-tack masking tape (i.e. drafting tape) over the graph lines, and to reveal them by stripping off the tape. If this method is used, the acetate sheet must be at least 150 microns (6 to 7 thousandths of an inch) thick. Some forms of flexible art tape or crepe tape allow quite complex curves to be dealt with in this way.

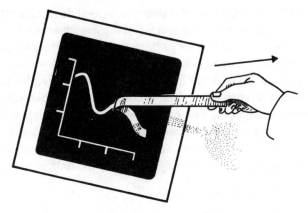

Figure 8.4 Animation of curves on a graph for overhead projection can sometimes be achieved by masking the line with 'low-tack' flexible tape which is wider than the curve. When the tape is pulled off the curve moves across the screen.

8.4. Testing OHP transparencies

As in the case of 35 mm slides, a practical test in the lecture theatre is the only reliable way of checking on the legibility of the transparencies and the success of the proposed techniques. The lecturer should practice using the overlays, reveals and masks, watched by a few colleagues sitting a reasonable distance from the screen. He should also take the opportunity of seeing the effects for himself by asking one of his colleagues to demonstrate them. Experience is the only real teacher, and it is not until the lecturer can see his work as the audience will see it that he will be able to learn from his successes and his mistakes.

9. Tape-slide programmes

9.1. Introduction

Existing literature deals with the subject of tape-slide programme making in detail. What appears here is a summary of the essentials for those who wish to become involved with this very effective method of teaching.

A tape-slide programme is not an illustrated lecture without the lecturer, but is specifically designed to impart information at a reader-paced speed. Tape-slide programmes have much in common with books in the sense that the audience is often only one person who views the display from a short distance. The images may contain detailed information, and can be projected for as long as the reader needs to absorb the content. He can pause wherever he wishes to do so in order to study the data or to make notes.

9.2. Planning a tape-slide programme

Tape-slide programmes need to be much more heavily illustrated than a lecture on the same subject. The absence of a lecturer removes some of the visual interest from the learning situation, and this vacuum must be filled to some extent by extra slides.

It is important to maintain a slide projection time of between 7 seconds minimum and 20 seconds maximum. This means that a new image should appear for approximately every forty words on the tape. The total length of the programme should not be longer than 20 minutes, and might contain a maximum of sixty to seventy slides.

It may be difficult for the designer to find enough illustrations to maintain the word to slide density ratio at about the correct figure (one slide = 20 seconds = 40 words). There is a very simple way of overcoming this problem which works especially well where objects are being talked about, whether it be patients, laboratory apparatus, or the use of techniques involving objects, etc. The trick is to present several slides of the same object or technique, but viewed from different angles. Even statistical data can be shown in this way. One slide might show the complete graph, the next might highlight a particular crisis point, the following one might enlarge a significant word or phrase in the subject matter, and finally the complete graph might be presented again. Four slides might therefore take the place of the one which would be used in a lecture situation.

As tape-slide programmes are reader-paced, the intended viewing time of 20 minutes may be prolonged to an hour or more, depending on how many times the reader wishes to repeat sections or pause to make notes. A 20 minute programme can therefore represent a 60 minute study period. Longer study periods will tend to be counter-productive however.

9.3. Writing the script

The script must be designed to be *spoken*. This requires a style of writing which is quite different from that typically used for published papers and books. The use of jargon and unfamiliar words should be minimised. Where such words are used, they may require slight emphasis and should be spoken slowly and pronounced with extra care. An accompanying slide spelling out any unfamiliar words will enable the student to make a note of their spelling and meaning.

The script must contain instructions for the speaker with respect to voice modulations, stresses and pauses. Particular note must be made of any pauses necessary during slide changes. These should sound natural, and should therefore occur at the end of a phrase or sentence. If the script is typewritten with triple spacing between lines and several character spaces at the end of each sentence, this will give adequate space for the insertion of instructional marks for the benefit of the speaker. These can be colour coded in some way.

Each page of the typewritten script should end with a complete sentence, or preferably a complete paragraph. The speaker can then pass to the next page during a naturally ensuing pause. Each page should be typed on one side only, and should be numbered very boldly so that the numbers can be seen 'out of the corner of the eye'. The sheets should not be stapled or fixed together in any way.

In reading the script, the speaker must retain an alert and interesting sound to his voice, with neither too much modulation (soft and loud passages) nor too little. For this reason, professional actors are often asked to speak the commentary. An amateur speaker will be greatly assisted by a script which has voice modulations, stresses and pauses clearly indicated on it.

Just as several drafts of a published paper may be necessary before the text is finalised, so most scripts for tape-slide programmes will require several recording and playback rehearsals before finalisation. A script planner (Figure 9.1) can help maintain a good balance between the commentary and the illustrations, and should be used from the beginning of the project.

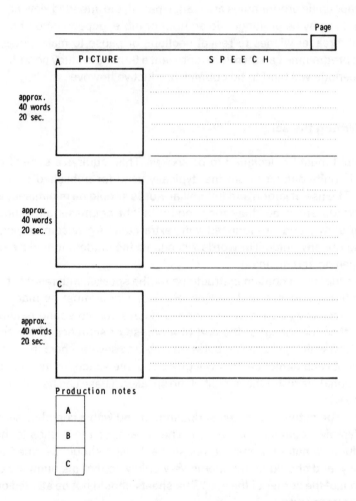

Figure 9.1 An example of a planning sheet for tape-slide programmes. Each large box represents a projected picture. The accompanying lines, if filled in with typewritten commentary, provide a natural limit to the amount of text before a new picture is needed. The full size of each sheet is A4. One page approximates to one minute of normal viewing time. These sheets are intended as a preliminary guide to programme production.

130

9.4. Graphic standards

Minimum line thickness and capital letter height may be less than those given for 35 mm slides as used for lecture purposes. Many tape-slide projectors have a screen size of 25 × 25 cm (10 × 10 in). If there is only one observer, he or she may sit so close to the screen that the reading distance is the same as that for a book. If, on the other hand, several persons are likely to view a programme together, which has the advantage of encouraging discussion, then their reading distance will be greater and line thicknesses and lettering sizes will need to be correspondingly increased. The designer of a tape-slide programme has no control over the number of observers, so it is safer to use the standards recommended for 35 mm slides.

A word of caution must be given in relation to the size of the image area on the slide. Some tape-slide projectors have not been designed to project the entire slide, but have a certain cut-off factor at the margins of the screen. This can be as much as 20% of the long dimension of the slide, and care must be taken when using existing slides which have data near the slide margins.

Tape-slide illustrations should be as attractive as possible, and should be consistent within a programme in terms of colouring and sizes and styles of lettering.

9.5. The recording

There are several points to observe here. The microphone should never be placed on a hard-topped table or shiny surface, otherwise unwanted sounds or sound reflections may be recorded. The table should be covered with a blanket and the microphone placed on an extra pad of semi-soft material such as expanded polystyrene, rigid latex or corrugated cardboard.

When reading the script the used sheet should be dropped onto a carpeted floor or disposed of in a manner which will not make a sound audible in the recording.

Recording should not be carried out in the centre of a room because equidistant walls can overlay recorded sound with unwanted echoes. Any other hard surfaces can have the same effect, so tests should be made to find the best position in the space available. Too many heavy curtains or absorbent surfaces, on the other hand, may deaden the recording to the point where it sounds flat and dull.

10. Television

10.1. Introduction

The television camera is not 'photographic', and the completely different technology of electronic scanning imposes very different standards on the designer. A helpful book on the subject is listed under 'Useful reference works'. What follows is of a general nature.

10.2. Graphic standards for television

10.2.1. Format

The transmitted television picture occupies an area 3 units by 4 units with the long dimension horizontal. The artwork must be planned to fit this area. There are, however, certain inherent inaccuracies in scanning, transmission and reception which affect the edges of the image area and which must be taken into account. A system of 'safe' borders has therefore been created for the designer to work within.

10.2.2. Size of drafts and artwork

The most economic size for artwork is 23 × 31 cm (9 × 12 in). Figure 10.1 shows the margins required for handling, the area which backgrounds should cover, and the 'safe' limits for important data.

10.2.3. Line thickness

On artwork of the size suggested above, the minimum line thickness should not be less than 1 mm. Alternatively, the minimum thickness can be calculated from the formula:

Figure 10.1A   The main elements of the composition must fall within the bold black line; any essential data such as words, etc., must be kept within the grey area.

Backgrounds which are intended to 'bleed' beyond the edges of the picture as seen on the receiver should be drawn within the dotted outer frame. Finally a plain area of card is left for handling in the studio.

Figure 10.1B   If standard 35 mm slides are projected for television, the transmitted picture area may cover only the grey zone, and details at the edges of the slide will be lost.

$$\frac{\text{Longest dimension of artwork}}{250}$$

10.2.4. Character height

The minimum capital letter height on artwork of the standard size should not be less than 10 mm ($\frac{1}{2}$ in), or:

$$\frac{\text{Longest dimension of artwork}}{30}$$

Choice of lettering style should follow the principles suggested for slides in Section 6.4.6. A simple undecorated 'open' typeface will give the best results.

10.2.5. Tones and shading

Cross-hatching and the use of closely set patterns of dots or lines can cause scanning interference and should therefore be avoided.

10.2.6. Contrast

Direct television scanning requires that artwork should be prepared within a restricted range of tones using neither black nor white, whereas with artwork for printed publications or slides, maximum contrast is essential. For television work, the reflectance (or tone) should range between 6 and 60%.

Intermediate tones need to be in definite steps, and for the designer the nearest equivalent can be found in the black and white 'screens' used by printers. These are made up of various concentrations of very small black dots, and they appear to the eye as different shades of grey. Figure 10.2 shows a convenient range of greys which will be distinct from one another when televised. Grey tones are best made by mixing black and white pigments until they match the scales shown.

10.2.7. The use of colour

Colours for direct television scanning need to be desaturated, which is again

134

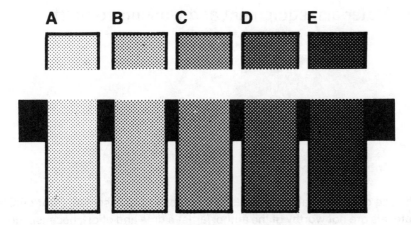

Figure 10.2 Contrast has to be restricted when making artwork for television. *A* will be seen as white, *E* as black.

different from the requirements for slides. Desaturation can be achieved by the addition of a little white to most colours. Pure, strong colours such as bright red can cause unwanted electronic interference patterns, or be converted into disturbing background noise.

However, indirect television scanning, such as telecine and machines which transmit data from standard 35 mm slides, may render invalid these recommendations on colour and contrast. It is therefore necessary for the author/artist to do the necessary research into local television systems first, so that the artwork will be compatible with them.

11. Materials, equipment and working comfort

11.1. Introduction

The use of good quality materials and equipment is essential. Poor quality materials are not worthy of the author/artist's time and effort, because more often than not they will result in disappointment and frustration. Good quality materials, on the other hand, will enhance one's skill and encourage further efforts.

A summary of all the materials and equipment which might possibly be needed is given in this chapter. Not all of these, however, will be necessary for the author/artist who is concerned mainly with simple black-and-white artwork for printed publications or slides. There are certain basic essentials which will be needed by anyone undertaking the preparation of his or her own illustrations, but beyond this, requirements will depend on how frequently illustrations are prepared, how complex they are and whether or not colour is used. The materials and equipment described below might therefore be grouped as follows:

Basic essentials for black-and-white artwork

1. Paper
2. Pencils
3. Inks
4. Pens
5. Rulers and set squares
6. Templates and stencils
7. Erasers
8. Scalpels
9. Burnishers
10. Adhesives
11. Drafting tape
12. Drawing boards

Other useful items for black-and-white artwork

1. Self-adhesive tapes (for black, coloured or patterned lines)
2. Dry-transfer symbols and lettering
3. Tone sheets
4. 'Pounce' powder
5. Fixatives and varnishes

Additional items for colour and OHP work

1. Self-adhesive colour sheets
2. Paints
3. Brushes
4. Cells (acetate sheets)

11.2. Basic essentials for black-and-white artwork

11.2.1. Paper

Good quality paper is essential in order to achieve a high contrast original. Many so-called white papers are not in fact white, as may be seen by comparison with printers' proofing paper. The surface of the paper is also important, since fine lines will only reproduce well if they are pure black and have smooth edges. An ink line drawn on bank paper, when examined with a magnifying lens, will be seen to have very rough edges (Figure 11.1). Although these lines may appear adequate on the original, they are likely to break up when reproduced by means of a high contrast photographic process. This can only be avoided by the use of good quality paper.

The best papers for artwork are not usually obtainable from stationery stores nor from the majority of shops selling artists' materials. The most likely place to obtain good quality paper is from a printer or a specialist graphic arts supplier. Proofing paper, Bristol Board or CS10 are recommended for use by the author/artist. Good quality artwork cannot be produced by the amateur on the cartridge, bond or bank papers which are commonly sold in stationery stores. Although tracing paper and films usually have hard surfaces and produce clean, smooth lines, these are not suitable for the amateur either, and should be particularly avoided.

The important characteristics of the various kinds of paper are summarised below.

Proofing paper (also known as proofing chromo or Baryta paper) is one of the best papers for the author/artist because it is the purest white available

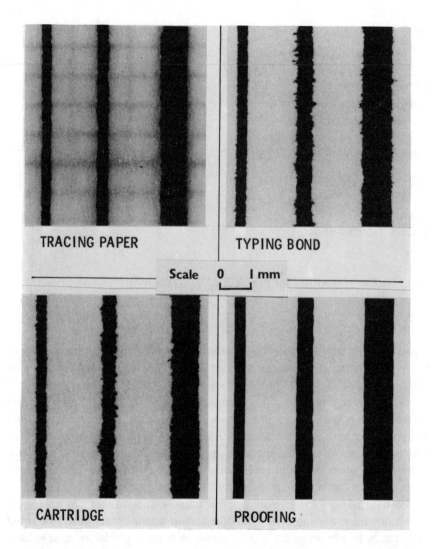

TRACING PAPER

TYPING BOND

Scale 0 I mm

CARTRIDGE

PROOFING

Figure 11.1A A photomicrograph of ink lines drawn on different papers. Proofing paper produces the best results – a smooth-edged black line with no reflectance from the ink. The same high quality technical drawing pens were used for each of the line weights (0.35, 0.5 and 0.7 mm respectively).

and it produces clean-edged ink lines. Usually only one side is treated for use, and this has a clay coating which may be either glossy or matt. The surface can be scraped with a scalpel blade without damage, and it will also accept typewritten images.

Bristol board and CS10 are also highly suitable for use. The best paper of this kind is made by Schoellerhammer. These are rag-based super-

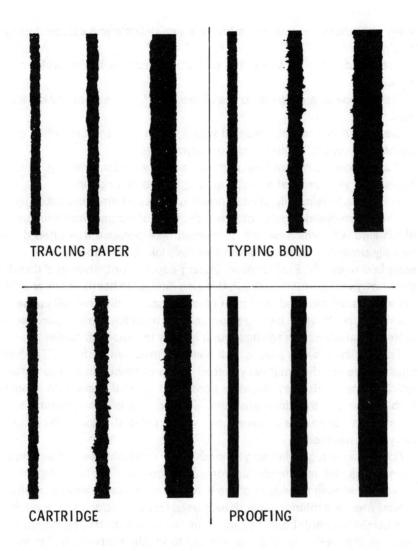

TRACING PAPER   TYPING BOND

CARTRIDGE   PROOFING

Figure 11.1B In this example, high-contrast printing techniques have been used to eliminate the difference in greyness of the various papers – all are now seen as white. Although the highlights from the tracing paper have been reduced, they have not been eliminated.

calendered papers with a hard, smooth surface on both sides. The surface is harder than that of proofing paper, and the texture is consistent throughout the thickness of the paper. This means that erasures and corrections made by scraping can subsequently be drawn over without difficulty. These papers do not accept typewritten images however.

Scraper boards are kaolin surfaced boards. They are not recommended by

139

many publishers because the surface is very delicate and can be damaged easily.

'Art' boards often have a grain, and are suitable for line or water colour work.

Fashion boards are similar to the above, but have a smooth finish for ink work.

Cartridge paper can be obtained with a rough or a smooth finish, and is suitable for dry work with pencils or crayons, etc.

Water colour papers are 'rag' papers made to withstand wetting. They are available in various finishes such as rough, smooth or grained.

Tracing paper is mainly used in drawing offices where transmitted light is used to make dyeline copies of drawings. It is not recommended for artwork of the kind with which we are concerned here because it is often dimensionally unstable, does not always lie flat, ink is slow to dry on it and is absorbed unevenly. Furthermore, tracing paper is not white and therefore does not give maximum contrast. It is also difficult to handle if cut-and-stick corrections are required; any form of water-based adhesive will cause it to buckle. Thus although tracing paper may be used for specific purposes by professional artists, for the amateur it is likely to cause difficulties.

Tracing film is a thin plastic with specially prepared surfaces. The better quality films are dimensionally stable. Some of these films have surfaces which damage ordinary drawing pens, and special pens with tungsten carbide tips are therefore available. Tracing film is not recommmended where high contrast is required, nor where cut-and-stick corrections are likely to be necessary.

Bond paper is a good quality smooth paper normally used for writing and typing. It is not recommended for quality ink work or fine line drawing however, because the edges of the lines will not be sufficiently smooth.

Bank paper is a thinner paper usually used for carbon copies in typewriting. It is semi-transparent and has much in common with tracing paper. Bank paper is extremely useful as an overlay to finished artwork or for making preliminary drafts and sketches.

Coloured paper is available in a variety of forms which are suitable for artwork. The cheaper papers are coloured by means of a chemical dye added to the pulp, and the colour may not be very stable. The surfaces of these papers are often rough and slightly fluffy, and artwork will be easier to execute if the entire surface is sprayed first with charcoal fixative. The more expensive coloured papers are made by printing colour onto one surface only.

Graph paper is obtainable in many different forms. The most commonly used grids are arithmetic or logarithmic, but there are others. Different scales are also available. These grids are printed in various coloured inks, blue, green, grey and orange being the most commonly used. When graph paper is used on machines which produce traces (such as electrocardiograms) the choice of printing ink may have an important bearing on the suitability of the trace for high contrast photography. Blue or green grids will disappear, orange grids will photograph as black, and grey grids may or may not photograph, according to the technique being used. Graph paper is usually printed on bond, bank or cartridge paper, but rarely on top quality paper.

Blotting paper should always be of good quality. The best kinds absorb wet ink immediately without causing it to spread first, and they do not leave fluffy debris on the drawing surface. It is advisable to select tinted blotting papers as these are easier to identify quickly in the working situation.

11.2.2. Pencils

Graphite (lead) pencils are generally unsuitable for high contrast line reproduction because they do not produce a black image and they tend to reflect light. If pencil lines are left unerased, however, they may appear on the final photographic print or projection slide. They should therefore not be used in any way that is difficult to erase. Pencil marks are sometimes difficult to erase from proofing papers. Graphite pencils are available in several grades, from hard to soft. Soft pencils (grades B or 2B) are usually easier to erase and are less likely to indent the surface of the paper. For marking the backs of photographs, pencils not harder than 2B should be used.

Charcoal pencils produce a good non-reflecting black, and are also available in a full range of colours.

Wax pencils are available in black and colours, and are used for working on smooth, shiny surfaces such as glass or plastic.

Crayons form a large group of pencils in which the pigment is in a water soluble, spirit soluble or oil soluble medium.

Non-reproducing pencils are usually pale turquoise blue in colour and the pigment is in a crayon or wax base. These pencils are used in the preparation of artwork for offset printing, and they do not normally record in high contrast photographic processes. Most blue and green crayons will have this property. Guidelines made in light blue/green therefore do not require erasing, which saves time and leaves the paper surface undamaged. On some surfaces, however, ink lines which cross blue pencil marks may be broken by the wax medium of the pencil. Blue lines should therefore be kept very light and fine.

11.2.3 Inks

It is essential that good quality black ink should be used on black-and-white artwork. Many so-called black inks are in fact dark blue, and not sufficiently dense to give a good photographic image.

Non-waterproof inks are the most suitable for use in technical drawing pens. These are available in black and colours. There are special varieties such as 'fount-india' for use in fountain pens. They can be diluted with water, and they are also suitable for use in airbrushes.

Waterproof inks are also available in black and colours. They are not recommended for use in technical drawing pens of the plunger and barrel type (see below), nor for fountain pens.

11.2.4. Pens

Technical drawing pens of the plunger and barrel sort are recommended, as these are designed to deliver controlled line thicknesses. The nib is basically a tube through which the ink flows, and in the centre of the tube is a wire which serves to control the flow of ink. In a good quality pen, the ink should flow freely but should not flood. Pens with stepped nibs should always be used, as these are less likely to cause ink to spread underneath rulers, set squares or templates (Figure 11.2). 'Staedtler' and 'Rotring' both manufacture pens of the desired quality. The micronorm series of sizes is advised.

Lettering pens may also be useful in some instances. The point of the nib is

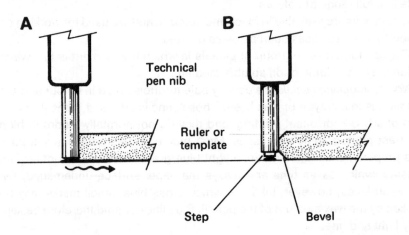

Figure 11.2 To avoid ink creeping underneath rulers or templates as in *A*, their edges should be bevelled and the pen nib should be stepped (all micronorm series pens have stepped nibs).

142

cut square or at an angle, depending on the style of lettering for which it is intended.

Sketching pens produce lines which vary in thickness according to the amount of pressure applied. Several grades are available, from hard to very flexible.

11.2.5. Rulers and set squares

The most useful rulers are made of plastic or stainless steel. Clear plastic rulers are best for preparing artwork, but should never be used to guide cutting tools. Straight steel edges are necessary for this.

Rulers and set squares should have 'square' rather than bevelled edges (see Figure 11.2). There is no advantage to bevelled edges when stepped-nib pens are used. Non-bevelled set squares can be used either side up without danger of over-riding guide rulers etc. Some of the micronorm stencils are made on a 'z' or stepped principle, and these are safer to use against square rather than bevelled edges. Plastic equipment of this kind should be at least 3 mm (1/8 in) thick.

Flexible rulers which can be bent to achieve smooth curves are also useful. There is a miniature variety avalable which can be used for very tight curves.

11.2.6. Templates and stencils

There is a large variety of plastic templates which can be used as guides for pencil or ink lines (Figure 11.4). French curves for drawing curved lines come into this category. There are also templates for circles, ovals, squares, triangles, flow chart symbols, chemical symbols, computer symbols, electronic symbols and many other subjects. Many styles of lettering guides or stencils are also available. For the scientific illustrator the best lettering stencils to choose are those in the micronorm series.

There are certain characteristics to look for when choosing templates and stencils. They must function well with ink pens, and be so designed that ink cannot flow underneath them. This can be achieved by several means. Some manufacturers supply an edging strip which raises the plastic from the drawing surface. So long as undue pressure is not put on the plate while drawing, this works well. The edging strips can be bought as a separate unit and fixed to almost any template. Other manufacturers bevel the edges of the openings forming the symbols, and the template can then be used on one side for pencil work and the other side for ink work. This type of template tends to be of thicker material, and this is an advantage.

Templates should be of clear coloured plastic rather than the colourless variety. The former are less easy to 'lose' among the other equipment being used.

11.2.7. Erasers

Rubber erasers are intended mainly for erasing graphite pencil marks. They dirty easily and cause dust, and they are liable to leave smudges and greasy marks, particularly on glossy proofing papers.

Vinyl plastic erasers are the most useful, as they will remove most lines (except ink) from a large number of different surfaces. They are less likely than rubber erasers to cause troublesome dust. A combined vinyl and ink eraser is illustrated in Figure 11.4.

'Art' cleaners are made from either natural rubber or vinyl, and they are intended to remove smudges and general dirt from drawing surfaces.

Ink erasers consist of natural rubber and abrasive materials. They can be used to erase ink-ribbon typewriter images, but they are not so effective with carbon-ribbon images.

Tracing film erasers are specially made to erase ink from tracing film. These contain microglobules of solvent.

11.2.8. Scalpels

A surgeon's scalpel is an essential piece of equipment, both for making corrections to ink work (see Section 12.3.4) and for the application of tone and colour sheets (see Section 12.5). The blades are detachable from the handles, and it is therefore possible to use different blades according to the task in hand. Several of the most useful blades are illustrated in Figure 11.3. Blades No. 15 and 23 are preferred. No. 11 is apt to dig holes in the paper when used for scraping, but can be useful for the accurate cutting of self-adhesive sheets. No. 23 is probably the most versatile blade. Its gently curving cutting edge gives a more easily controlled scraping action.

11.2.9. Burnishers

Burnishing is a technique used for restoring the paper surface after ink work

Figure 11.3 Scalpels *A* and *B* are the choice for most correcting and scraping-back procedures. The amateur will probably find blade 23 the most useful. *C* has advantages when cutting out intricate shapes in self-adhesive film, as the narrow handle makes twisting and turning easy.

A

C

B

Figure 11.4A-C
  A. Drafting tape (sometimes called masking tape). Can be used to fix drawing to board, etc.
  B. This eraser made by Staedtler, Germany, is in two parts. The light-coloured part erases pencil without making dust and the dark part contains an abrasive for removing ink and more stubborn marks.
  C. Pounce powder marketed by West & Partners. Useful for cleaning paper, film, glass, etc. before ink work.

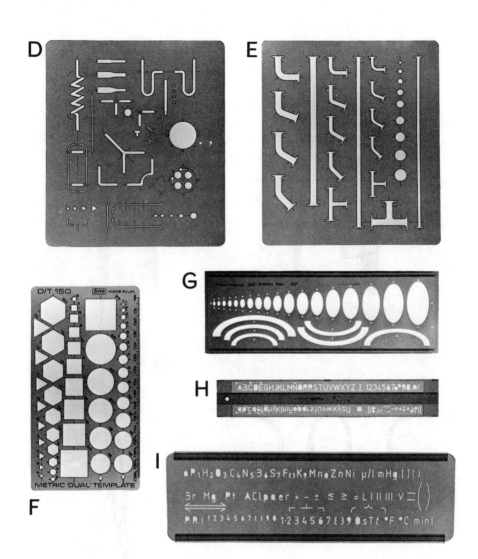

Figure 11.4D-I

D. 'Uno-lab' template made by West & Partners, UK. Specially made for laboratory diagrams.

E. 'Uno-lab' template made by West & Partners, UK. Designed for drawing laboratory tubing of different diameters, including various joints and junctions.

F. A useful general-purpose template made by West & Partners. Particularly useful in preparing organic chemistry drawings.

G. Oval templates made by Minerva, France, provide the most convenient arrangement of large and small ovals.

H. Good letttering templates made by Rotring, Germany. These are almost unbreakable and are so designed that the lettering being drawn cannot be smudged when the template is moved.

I. 'Uno-lab' template made by West & Partners, UK. Designed for use with 0.5 mm diameter technical pens of the micronorm series. This template has been designed for use in organic chemistry.

147

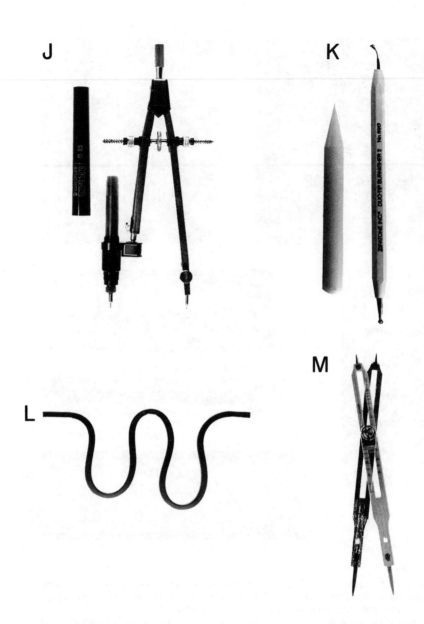

Figure 11.4J-M

J. Strong compass with fine adjustment for drawing pencil or ink circles using technical pens. Made by Staedtler, Germany.

K. Burnishers used to rub down dry-transfer or self-adhesive products. The one on the left is made from a piece of nylon. Scale: burnisher on the right is 17 cm (6.5 in) long.

L. Flexicurve minor. An especially useful flexible ruler for drawing 'tight' curves.

M. Proportional dividers by Staedtler, Germany. These are well-made precision instruments, extremely useful for dividing lines or circles into equal parts without mathematics.

148

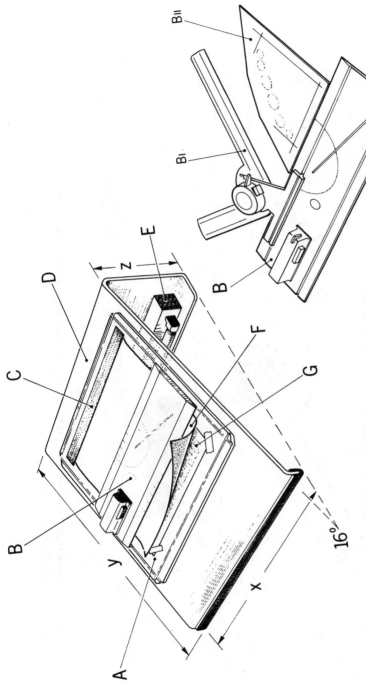

Figure 11.5A The Rotobord transilluminated drawing unit, with Marabu clear acrylic boards, rulers and set squares. The desk unit is shown here in 'drawing mode' (see Figure 11.4C for 'projection mode').

Desk-top drawing/projection unit designed by Doig Simmonds. A) clear acrylic drawing board; B) clear acrylic ruler which can be placed in a groove on any of the four sides of the board; C) magnetic clip holding paper; D) desk-top transilluminated unit; E) 8 watt 250 V 50 Hz flourescent tube.

x = 40 cm (15¾ in); y = 41 cm (16 in); z = 14 cm (5½ in).

149

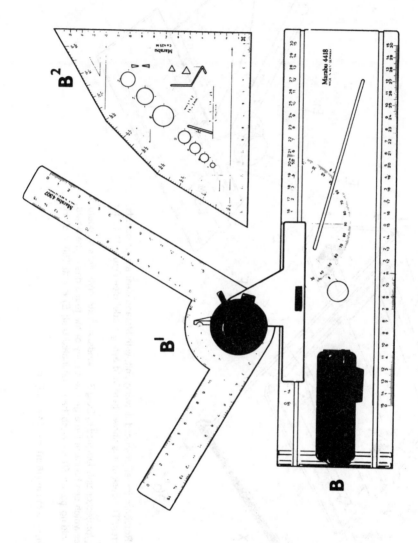

Figure 11.5B Details of B, B¹, and B². The well-machined and finely printed equipment, B¹, is adjustable through 90°.

plan

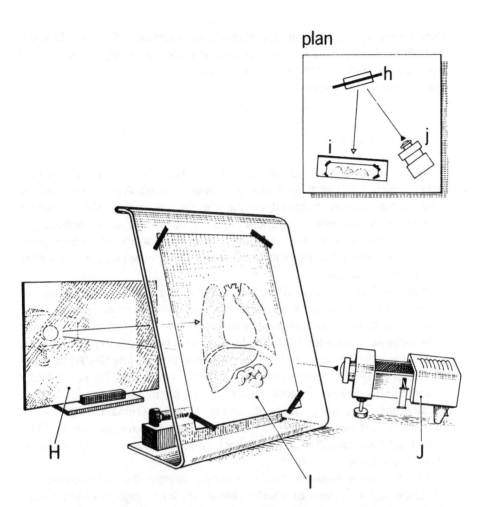

Figure 11.5C The desk-top unit in 'projection mode'. This system enables the operator to trace directly from projected slides. *H*) mirror; *I*) tracing paper; *J*) projector.

has been removed with a knife (Section 12.3.4), and for rubbing down dry-transfer materials and low-tack self-adhesive tone or colour sheets (Sections 12.4 and 12.5). Burnishers are often thought of as an unnecessary luxury, and many people try to make do with any suitable object which comes to hand, such as a ballpoint pen or pencil. These makeshift tools can cause problems, however, and their shape limits their use. Ink from ballpoints and graphite from pencils can cause unwanted smudges on the artwork, and their sharp points can easily crack dry-transfer images.

A properly designed burnisher (see Figure 11.4K) should have a tapered and slightly curved point which is not sharp. The flat of the tool can then be used for rubbing down lettering and tones, as described in Sections 12.4 and 12.5. Burnishers should be made of ivory or agate, since these materials will not mark the surface of the paper. A metal burnisher will mark the surface of proofing paper very easily.

11.2.10. Adhesives

Adhesive used for cut-and-stick corrections (see Section 12.3.2) can ruin the artwork unless carefully chosen and carefully used. Adhesives must dry white or transparent, must not discolour or stain the paper, should remain reasonably flexible when dry, and should not be too wet when applied.

Rubber solutions may cause staining after a period of time on some papers, and they may dry out and lose their adhesion. They are not recommended for non-professional use.

Wax based adhesives are usually in the form of aerosol sprays. They are quick and clean to use, and will not cause the paper to buckle. They are more suitable for fixing large patches rather than small ones however, and they may not be suitable when mounting heavy card or heavy photographic paper.

Resin emulsion adhesives, incorporating polyvinylacetate (PVA) or poly-vinylchloride (PVC), are the most useful kinds of adhesive. These are water based, but become waterproof when dry. They are milk-white in use but dry transparent. They are 'permanent', and most of them remain flexible. PVA and PVC emulsion adhesives are the safest, cleanest and easiest to use. The most useful varieties come in hard block form. A popular brand is made by Henkel, and is called 'Pritt'.

All emulsion adhesives should be used sparingly. It is not necessary to cover the entire surface of a patch, especially when high contrast photography is being used. If adhesive is used too freely the water in it may shrink or cockle the paper, thus causing the photographer unnecessary difficulty with lighting.

Dry mounting is a further method of sticking materials together. This requires a heated flat-bed press and shellac foil. The artwork must be able to withstand heat. This is the best method of mounting heavy weight photographic paper, though resin coated and plastic photographic papers need special techniques.

11.2.11. Drafting tape

Drafting tape is a self-adhesive tape with 'medium-tack' properties. Because of these properties it is easier to remove than Sellotape, for example, and it does less damage to paper surfaces. It can also be used several times before it loses its stickiness. The 2.5 mm (1 in) width of tape is generally the most useful (Figure 11.4).

Drafting tape is particularly suitable for sticking drafts to the drawing board or to the artwork (Section 13.3.1), for removing unwanted dry-transfer items (Section 12.4.3) and for removing surface debris from a drawing.

11.2.12. Drawing boards

Drawing boards are obtainable in a variety of forms and sizes. They may or may not have transilluminated panels, may or may not be adjustable, and may or may not have drawing 'heads' or parallel motion devices.

Drawing boards for technical illustration work should, first and foremost, have a flat surface. They should incorporate some form of transillumination, usually in the form of a panel illuminated from below, and they should be adjustable to a comfortable working angle. This will normally be an angle of 16-20° at table top height. A parallel motion or moving ruler system will be an advantage for drawing vertical, horizontal and oblique lines.

The size of the board is important. For the kind of illustrations executed by the author/artist (i.e. simple graphs, charts and diagrams), a large board is of no great advantage. This will be particularly true if the equipment needs to be folded up or tucked away in a cupboard when not in use. The best size, therefore, is one which suits the size of most original drawings. Since A4 is the standard size for artwork recommended here, then an A4 board may be all that is required. An A3 board may be preferable, however, as the larger working surface gives greater flexibility when using A4 paper vertically or at an angle.

Rotobord Ltd. now market a transilluminated drawing unit of A4 size specially designed for the author/artist who requires an efficient and well made instrument which can be used conveniently on an existing bench or table without occupying too much space (Figure 11.5A, B, C).

11.3. Other useful items for black-and-white artwork

11.3.1. Self-adhesive tapes

Self-adhesive tapes can be used to 'draw' lines (Figure 11.6). They are available in black, white and several colours, and in a variety of different

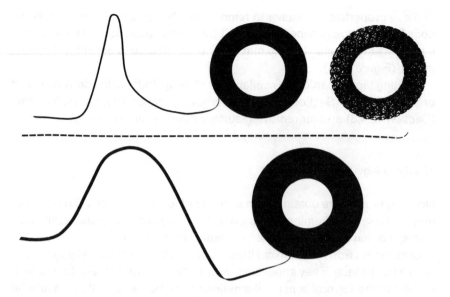

Figure 11.6   Tapes. Several manufacturers market a variety of plain and patterned self-adhesive tapes in different widths. Some of the plain tapes are flexible, and these can be used to produce quite 'sharp' curves.

widths. These widths are not yet fully metric and do not match the micronorm standard, but this will probably come in due course. Some of these tapes are flexible and can be used for curves and sharp corners. They are quick and easy to apply and make working with inks obsolete in some circumstances.

Tapes do have some disadvantages however. Crêpe-paper tapes should be avoided if the artwork is to be photographed as they produce highlights. If the artwork is subsequently copied by some kind of contact copying process, the thickness of the tape may prevent good contact. If the artwork is put through a typewriter for the addition of lettering, the rollers may dislodge the tape. All illustrations require careful planning in relation to the techniques to be employed if frustrations of this kind are to be avoided. Thus if an illustration is to have lettering typed on it, the application of the tape should be left until the typing is finished.

11.3.2. Dry-transfer symbols and lettering

The principle of these materials is that images are transferred from a waxed carrier sheet to the artwork. The system is suitable for any form of frequently repeated symbol or unit, such as lettering, numerals, chemical formulae,

154

logos, trade marks, etc. At present there are some eight manufacturers producing items which are of special interest to the scientific illustrator. The most useful and readily obtainable brands are 'Rapitype' made by Graphic Systems International, and 'Alfac' which is made in Belgium. 'Rapitype' is available in twelve different colours.

There are two different systems in use for creating dry-transfer images. One of these systems is photographic, and it is the photographic image itself which is transferred to the artwork. Very fine detail is possible with this system, but the image density is not as great as with the alternative system. This can be a disadvantage if the artwork is to be transilluminated, as in the case of X-rays for example. The alternative method is to create the images using a silk screen printing process. The images are not always sharp edged and very fine detail is not so easy to obtain, but the images have a greater density than those produced by the photographic method.

Dry-transfer symbols should release from their carrier sheet when light pressure is applied. They should not suffer from 'pre-release', i.e. detach themselves from the carrier when and where they are not wanted. The images should remain flexible, and they should not crack while being applied. Similarly they should be able to withstand the heat generated by some kinds of photocopying and xerographic copying machines without distortion or damage.

Prolonged storage or unsuitable storage conditions may lead to premature deterioration of the product. Photo-image materials often have a much stronger carrier sheet than the silk screen kind, and usually store well. Both systems, however, use waxes to secure the images to the carrier sheets and to ensure their adhesion to the artwork. These waxes may behave unpredictably if they have been subjected to unsuitable conditions.

The non-professional user is advised to purchase only those materials required in the short-term. Some manufacturers market smaller sheets, sometimes called 'quarter sheets', and these are usually much more economical than full size sheets. They also allow the user to purchase only those items which are actually required. Both of the manufacturers mentioned above make quarter or small-size sheets.

11.3.3. Tone sheets

Even a black-and-white illustration may have 'tones' added to it without destroying the essential economy of black-and-white reproduction techniques. These tones consist of a spread of dots, dashes and stipples. They may either be bonded to a self-adhesive carrier sheet which is cut up and applied to artwork, or they may be applied using the dry-transfer principle. In most cases, self-adhesive toning is quicker and easier to use than the dry-transfer variety as far as graphs, charts and diagrams are concerned. Self-adhesive tone sheets should be of the 'low-tack' variety, i.e. they should not stick firmly to the paper surface in such a way that they are difficult to remove or replace, until they have been burnished. The best of these is marketed under the trade name of 'Zipatone'. 'High-tack' sheets are difficult to handle, and may also accidentally remove ink lines or previously applied dry-transfer items.

Tones vary from the very coarse to the very fine, and the author/artist would do well to remember the standards mentioned earlier in this book in relation to tones for use on artwork which is to be reduced by 50% or which might be used to make slides. These standards state that dots should not be less than 0.5 mm in diameter and at least one dot diameter apart; lines should be not finer than 0.35 mm and at least 1.2 mm apart. Even within these limitations, however, there are many tones to choose from. Some of these are illustrated in Figure 5.4.

11.3.4. 'Pounce' powder

'Pounce' or pumice powder is a cleaning powder which can be used for the preparation of a large number of surfaces so that water based inks and colours can be used on them (Figure 11.4). This includes all white papers, tracing papers, films, cells, glass and plastic. The powder may be hard to remove from coloured paper however.

11.3.5. Fixatives and varnishes

Fixative is a colourless medium used to prevent pencil or charcoal work from smudging and to help protect water colours from damage and loss of colour. It can also be used to consolidate an otherwise 'fluffy' paper surface, thus making it more suitable for ink or paint work. Fixative always dries matt and may have a very slight darkening effect, though this is usually not a disadvantage. It does not provide a hard or fully waterproof coat as do some varnishes. It is best applied from an aerosol can, but can also be applied with a mouth spray.

Varnishes are usually applied from an aerosol can or with an air brush. It is usually not advisable to apply varnishes with a brush. There are several varieties of varnish available, giving an acrylic, vinyl, polyurethane or cellulose coating which can be either matt or glossy. Tests should be made prior to using a varnish, as serious and irreversible changes in colour and tone can occur with some materials.

11.4. Additional materials for colour and OHP work

11.4.1. Self-adhesive colour sheets

For coloured work, a full range of hues (at various levels of intensity and saturation) is available in the form of self-adhesive sheets. The colour must be translucent for OHP work.

11.4.2. Paints

There are two basic kinds of paints, i.e. water-soluble or oil-soluble paints.

Water-soluble paints include water colours, crayons, pencils, vinyl and acrylic paints, gouache and poster paints. Among the most useful of these are water colours and acrylic paints. Water colours should be used on water-colour paper for best results. Acrylic colours can be used on almost any surface, though some absorbent surfaces may require an acrylic undercoat. These colours are waterproof when dry and remain fully flexible without cracking.

Oil-soluble paints include oil paints, crayons and pencils. Oil paints usually require careful preparation of the ground before work commences, and in general work progresses less quickly with oil-based colours than with water-based colours.

11.4.3. Brushes

A large variety of brushes is available. They are usually made either of hair (sable, squirrel or ox bristle) or of nylon. Nylon brushes have the advantage that each individual 'hair' can be of a controlled shape, tapering consistently towards the working end. These brushes were designed principally for acrylic paints, but are good for all pigments.

Brushes may be round, oval or flat in shape. Round brushes should come to a distinct point when wet. Oval or flat brushes should form a fine, consistent edge without splitting when wet. The sides of the brush should taper slightly inwards.

11.4.4. Cells and foils (acetate sheets)

These are water clear films which are obtainable in a variety of thicknesses either in sheets or in rolls. They are used mainly for overhead projection work, cine film animation or special-effect slides. Cells may be plain or pre-treated for ink and pencil work. Plain cells can be prepared for water-soluble inks (or felt-tipped pens with water-soluble inks) by using 'pounce' powder.

11.5. Working comfort

11.5.1. Organisation

A well-organised work place or working area will speed production and reduce frustration. All essential materials and equipment should be arranged so that the minimum amount of searching or movement is necessary to find and pick up any item. Small tables or drawer units placed to the right and left of the working table may provide all the additional work surface required.

11.5.2. Furniture

No professional craftsman would ever tolerate the working environment which so many amateurs put up with. A comfortable work place is essential for professional results.

The best kind of seating is an adjustable typist's chair (without arm rests) on castors. The height of the seat should be about 45 cm (18 in), so that the feet touch the floor. The backrest should be adjusted so that it supports the small of the back. The castors will allow movement to and from adjacent work surfaces and storage units while sitting down.

The work surface should ideally be 68.5 cm (27 in) from the floor, rather than the usual 76 cm (30 in). The surface should be at least 1 m (40 in) wide by 60 cm (24 in) deep. This will be large enough to take an A3 drawing board. The drawing board should be set at an angle of about 16° from the horizontal. Maximum comfort is achieved when it is possible to draw while sitting back or leaning forward.

11.5.3. Lighting

Suitable lighting is extremely important for working comfort and for the production of high quality artwork. Some form of transillumination set into the drawing surface is the most essential requirement. Not only does this make it possible to see the draft or sketch through the drawing paper, but light coming through the paper itself eliminates shadows and helps to control unwanted reflection from plastic equipment.

The amount of overhead lighting can be reduced if the drawing surface is transilluminated. The position and angle of any additional light sources should be fully adjustable. The anglepoise lamps made for architects have longer arms than many models, and the bulb is better placed.

12. Basic techniques

12.1. Care and preparation of paper

Paper should be bought clean, kept clean and stored flat. Care must be taken when handling paper. The surface of better quality papers is easier to damage than that of the cheaper ones. Bristol Board and CS10 are particularly liable to acquire invisible finger marks which, being greasy, will prevent ink from adhering to the paper. These hidden grease marks can be removed with 'pounce' or pumice powder (see Section 11.3.4), or even with talcum powder. A small amount of powder is rubbed over the surface of the paper with cotton wool, using only gentle pressure. The softer proofing papers require the lightest pressure. It is often worth treating paper in this way before use, especially if it is known to have been handled frequently. Paper should never be cleaned with any water or spirit based fluids.

Visible marks can sometimes be removed with plastic erasers (see Section 11.2.7). Erasing should be carried out very gently, changing the direction of movement often so that the paper surface is not damaged. Heavy erasing in one direction may cause any subsequently drawn ink lines to show altered characteristics over the area which has been cleaned.

After any erasing procedure or after the use of pounce powder, the debris should be brushed away with a soft paint brush. A 5 cm (2 in) or 7.5 cm (3 in) home decorating paint brush is ideal for the purpose. Debris should never be wiped away with the hand because this is likely to make new invisible grease marks on the surface of the paper. Blowing onto the surface is also inadvisable as any fine drops of moisture can alter the surface texture irreparably.

12.2. Ink work

12.2.1. Use of technical drawing pens

The instructions provided by the manufacturers of technical drawing pens

160

describe their construction and maintenance in detail, and it is unnecessary to repeat this here. The essential feature of plunger-in-barrel pens is that they deliver ink to the paper via a tube, and this tube has a carefully machined diameter which precisely determines the width of the resulting ink line. This system only functions properly, however, when the tube is at right angles to the paper surface (Figure 12.1A). If the pen slopes, even slightly, then the line will vary in thickness and density and may also have ragged edges. It is therefore essential to learn to hold these pens at 90° to the paper surface, and

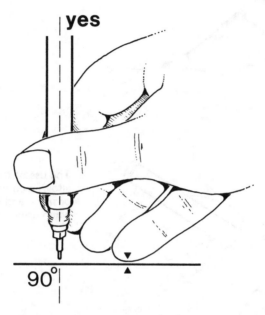

Figure 12.1A Using the technical drawing pen. The most vital requirement of all is to hold the pen correctly. To obtain a smooth ink line of consistent width, the pen must be held at right angles to the drawing surface. The little finger is sometimes useful (a) to prevent too much pressure on the nib and (b) as a guide to maintaining a vertical position.

not in the normal handwriting mode. The initial awkwardness of this task can be reduced by fixing an angle piece to the nib as shown in Figure 12.1B. Most manufacturers supply an angle piece with their pens when they are bought as a set.

The pressure of the pen on the paper should be light but positive, i.e. about the same pressure that would be used when drawing with a soft pencil. It is advisable to experiment with different pen pressures on the same paper which will be used for the artwork before beginning work on the drawing itself. It is also important to draw slowly and without variation in speed.

Pens should be kept clean during use by wiping the nib frequently with a

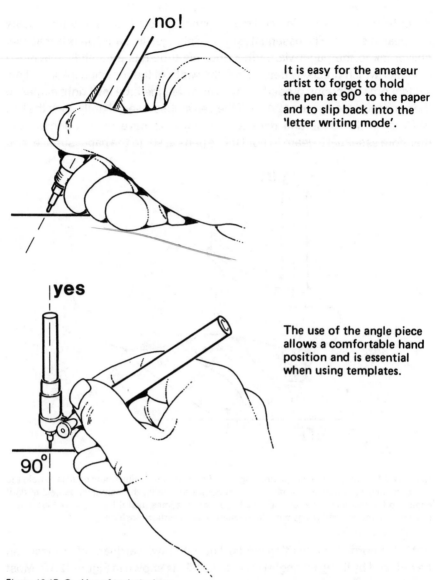

no!

It is easy for the amateur
artist to forget to hold
the pen at 90° to the paper
and to slip back into the
'letter writing mode'.

yes

The use of the angle piece
allows a comfortable hand
position and is essential
when using templates.

90°

Figure 12.1B,C Use of technical pens.

piece of paper tissue or lint-free cloth. This will help to prevent debris from
the paper surface from clogging the ink tube and causing an erratic flow of ink
or a total blockage.

Pens should always be capped when not in use, or alternatively they may
be parked in a humidifying rack (Figure 12.2). These racks are sometimes
supplied as part of the pen case, or the cap itself is used as the humidifier.

162

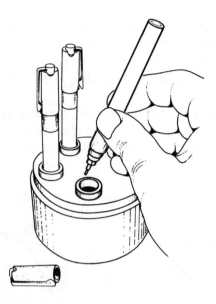

Figure 12.2 Special holders are available for the storage of pens or for 'parking' them during use.

A humidifying system keeps the ink from drying in the nib.

NEVER leave the pen lying about with nib uncapped when it is not being used.

12.2.2. Blotting

A piece of good quality blotting paper should be kept handy at all times. Some artists like to rest the hand on a piece while drawing. This not only protects the underlying surface from grease and perspiration, but the blotting paper is also available for immediate use. Ink lines drawn on good quality proofing paper may not need blotting at all, because the ink is absorbed quickly. If there is any excess ink, and there may be some at the beginning and end of a line, blotting can usually be done immediately after the line has been drawn without reducing its blackness. Tracing papers and films absorb ink less rapidly however, and blotting should be avoided because it will reduce the density of the image. Drawings on these kinds of materials may therefore take longer to complete than those on other papers.

12.2.3. Drawing ink lines

Straight lines can obviously be produced using a ruler and, if necessary, a set square (see Section 11.2.5).

163

Oblique lines can be accurately drawn with adjustable set squares or by means of special devices which can be attached to the drawing board.

Curved lines can be produced with 'french curves' or flexible rulers.

Circles can be drawn with plastic templates, or by using beam or axial compasses.

Lines, spaces or circles can be divided using 'proportional' dividers, but these are expensive items. A cheaper method is to use the scales illustrated in Figure 12.3 which provide for the division of a circle into polygons, angles or percentages.

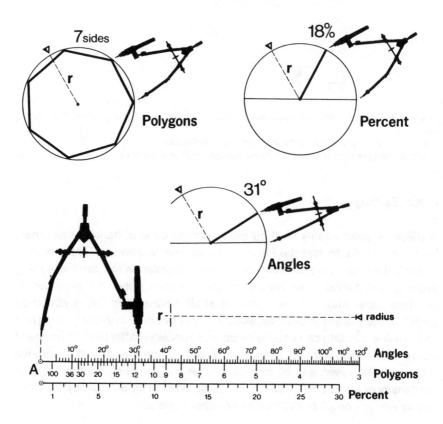

Figure 12.3 The drawing on this page can be used to make pie charts (by dividing circles into percentage sections) and polygons, and to create angles.

1. Draw a circle of the radius shown.

2. Put the point of the dividers or compass at 'A' and select the item required on the appropriate scale. An example of how to draw a 31° angle is illustrated.

(Adapted from data supplied by courtesy of Staedtler GMBH, Germany.)

12.2.4. Finishing-off ink lines

Ink lines drawn with a technical pen have rounded ends, and the corners of rectangular shapes tend to be rounded . This is particularly noticeable with thicker lines. For a professional finish to a drawing, these round ended lines and corners should be squared off with a scalpel (Figure 12.4). This will be easier if the lines are drawn slightly longer than required and if one of the lines making a corner is longer than the other. When the ink is completely dry, the lines can be trimmed back with a scalpel. This may sound time-consuming, but in fact it will take only a few minutes; it is this level of attention to detail which makes the difference between amateur and professional work.

12.2.5. Use of templates and stencils

When used for ink drawing, templates and stencils should be fitted with an edging strip if the openings forming the symbols are not bevelled. This will prevent ink from flowing under the edges. Micronorm stencils are designed to avoid smudging when moved over freshly drawn lettering however (see Section 11.2.6).

Templates and stencils may scratch the drawing surface and damage existing ink work when they are slid along the edge of a ruler or set square. The drawing can be protected by placing a piece of typewriter copy paper under the ruler with an extra 5-10 mm ($\frac{1}{4}$-$\frac{1}{2}$ in) projecting under the path of the template or stencil (Figure 12.5). The main body of the instrument will then slide on the copy paper instead of on the drawing surface.

Stencilled lettering may seem difficult to execute satisfactorily at first, but it is relatively quick and easy after a little practice and costs nothing. Particular care should be taken over the spacing of the lettering. The majority of letters are already cut the correct distance apart on micronorm series stencils, but some capital letters may need slightly more or less space according to which letter is next to which (see Section 5.2.6). A spacing square is usually provided on lettering stencils to give the correct space between words. If the top edge of the stencil is lined up with the lettering already written, the new line will then be the correct distance from the one above.

12.2.6. Short cuts

Scale calibrations on graph axes, standard error finials, short cross bars and other similar items are time-consuming to draw and often difficult to do

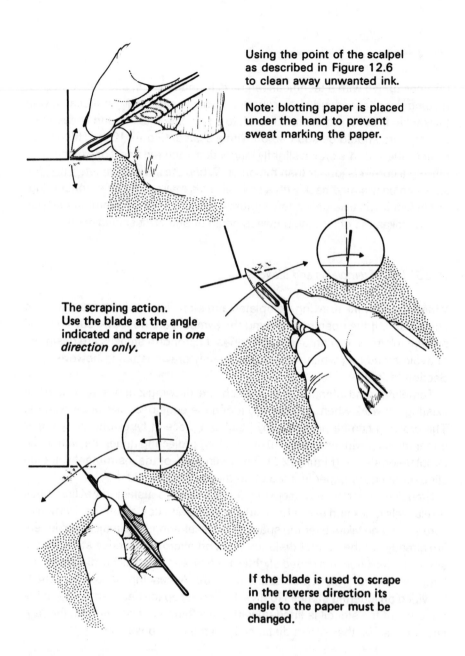

Using the point of the scalpel as described in Figure 12.6 to clean away unwanted ink.

Note: blotting paper is placed under the hand to prevent sweat marking the paper.

The scraping action. Use the blade at the angle indicated and scrape in *one direction only*.

If the blade is used to scrape in the reverse direction its angle to the paper must be changed.

Figure 12.4 Use of scalpels.

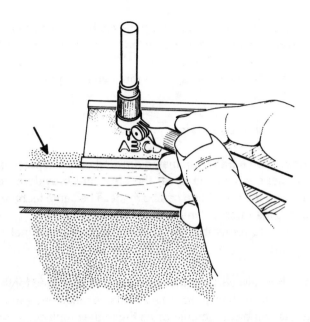

Figure 12.5 When using other equipment such as templates, rulers, etc., thin bank paper is placed over the drawing for the template to slide on (as indicated by the arrow).

neatly, particularly if the ends of each line are squared off as recommended above. Much time and trouble can be saved and a more professional result achieved if appropriate dry-transfer dashes and ruled lines are used instead. Sheets with dashed lines can be obtained from many dry-transfer manufacturers, and some are compatible with the micronorm series of standard line thicknesses. Short dashes about 3 mm long and with a line thickness of 0.35 or 0.5 mm make excellent standard error finials. Dashes of the same length or a little longer can be used for scale calibrations. (A dash can also be used to convert a typewritten or stencilled 'u' into a 'mu'.)

12.3. Correcting errors in ink work

12.3.1. General comments

Errors are usually not the tragedy that they may appear to be at first, and it is usually better to leave them uncorrected until all the ink work is finished. It is, however, essential to make any necessary corrections and to 'clean up' the drawing before applying any tapes or tone sheets.

Providing that high contrast photography is being used to reproduce the artwork, then any major corrections can be made by 'cut-and-stick' methods. For smaller corrections, Bristol board, CS10 and proofing paper can be scraped clean with a sharp scalpel without damaging the drawing surface.

12.3.2. Cut-and-stick corrections

With the artwork on a transilluminated drawing board, a patch of paper can be cut to a shape which will cover the error and provide a new drawing surface on which the correction can be made. The same technique can also be used for the insertion of typewritten or typeset lettering.

It is important to remember the following points when making cut-and-stick corrections:

1. The patch must always be of the same paper as that used for the rest of the drawing, so that the background has the same reflectance and the ink lines will have the same characteristics as those they replace.
2. Whenever possible, the patch should be cut so that re-drawn ink lines do not need to cross it edges. Edge-crossing is a technique in itself, and however skillfully it is done the crossing points may be visible in any subsequent photographic prints.
3. Patches should always be trimmed as close as possible to the boundaries of the new artwork. Unnecessarily large patches may prevent proper contact if the drawing is to be copied subsequently by a relfex or contact copying process (see Section 5.4.2).
4. When fixing small patches to artwork, the adhesive should always be applied to the patch and not to the artwork. This eliminates the risk of adhesive being applied in the wrong place.
5. Beware of excess adhesive around the edges of patches, as this will pick up dirt.
6. Small patches can be impaled on the point of a scalpel as an aid to accurate positioning.

12.3.3. Edge-crossing

If it is impossible to avoid the situation where new lines cross from a patch to the original drawing surface, then the following points are important:

1. The edge to be crossed must be stuck down firmly, but adhesive must not

be allowed to exude. If it does, it should be removed with a scalpel. Ink lines drawn over adhesive will either spread or shrink, depending on the type of adhesive.

2. The new line should be drawn steadily and slowly from the patch outwards. When the edge is crossed, the pen nib will then step down rather than up. The latter direction of movement is likely to result in the nib catching on the edge of the patch, which would inevitably cause a blemish.

3. If a small gap in the line remains at the step, some touching-up will be necessary. This can be done with a finer pen than that used for drawing the line. If, on the other hand, a blob of ink appears, this can be trimmed down to the correct width with a scalpel. The ink must be allowed to dry properly first however.

Patching on tracing paper or film is doomed to failure because the patches alter the reflectance value of the basic material. An additional difficulty is caused by the dimensional instability of many tracing papers, which tend to buckle when adhesive is applied to them. Tracing papers are therefore not recommended (see Section 11.2.1).

12.3.4. Correcting with a knife

Ink lines can be reduced in width, eliminated or 'shaded off' using a scalpel, providing the artwork is on good quality paper. The blade should be held at about 90° to the paper surface as shown in Figure 12.6, and a gentle scraping action should be used. The sharper the scalpel the lighter the touch required and the less damage will be done to the surface of the paper. A blunt blade requiring heavy pressure will dig up the paper surface, thus ruining any subsequent ink work. A sharp blade will make a clear ringing sound on the paper surface, whereas a blunt blade will make a much duller sound.

On good quality paper, any damage caused by scraping the surface can be partially rectified by burnishing the surface before applying new ink lines. The artwork must always be placed on a hard surface to prevent the paper from dimpling as a result of the pressure applied.

12.3.5. Correcting with paint

White water-based paints (such as poster colour or 'process white') and correcting fluids for typists may be used to eliminate errors, but there are dangers in this method. Firstly, the paint may crack and peel off as a result

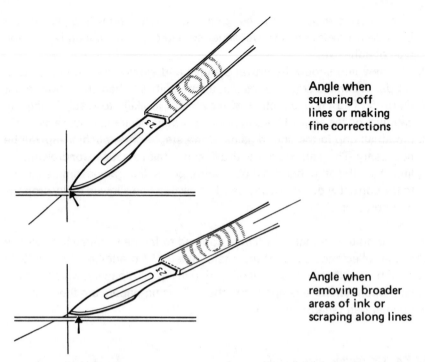

Angle when
squaring off
lines or making
fine corrections

Angle when
removing broader
areas of ink or
scraping along lines

Figure 12.6 Corrections with the knife. A *light* touch and a *sharp* scalpel are essential for successful correction.
REMOVE INK ONLY, *NOT* PAPER

of the handling which the artwork will receive before it reaches the printer and is photographed. Secondly, if the paint is too generously applied, the surface may become raised. If the raised area collects dirt, the white correction will then photograph as black. If paint is used on areas where tone sheets will be subsequently applied, the raised surface may cause difficulty in photography.

Paint can be useful, however, if sparingly applied with a fine brush. It should not be too wet or the paper will cockle, thus creating difficulties for the photographer. For this reason, paint is only suitable for small corrections.

12.4. Use of dry-transfer materials

12.4.1. Lettering

The general remarks made in relation to the spacing of stencilled lettering also apply here.

170

Dry-transfer images are applied in two stages. First of all, the 'point' of the burnisher is used with *light* pressure to effect the exchange of the image from the carrier sheet to the artwork (Figure 12.7A). The use of light pressure only will ensure that the image does not crack. The 'flat' of the tool is then used to firmly burnish the image onto the artwork (Figure 12.7B). A piece of carrier paper or bank paper should always be placed over the artwork during the final burnishing. This will prevent friction damage.

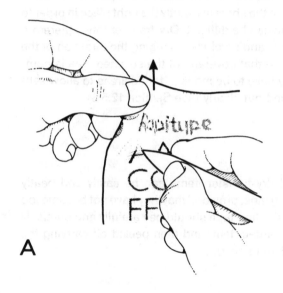

Place the sheet of lettering or symbols in the desired position over a guide line. Burnish *lightly* and evenly across the letter. Meanwhile, pull the carrier sheet upward slightly against the point of the burnisher.
This is not work to be done in a hurry.

A

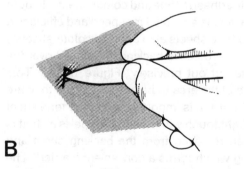

After the letter has been transferred, use a backing sheet under the burnisher and rub rigorously.

B

Figure 12.7 Stages in applying dry-transfer lettering or symbols.

12.4.2. Tones

Dry-transfer tones should ideally be applied on a transilluminated drawing surface so that the areas to receive the tone are clearly visible. Those areas which do not need to be tinted should be masked with typewriter copy paper (bank paper) to protect them from the accidental transfer of tones.

Tones are applied by burnishing, in the same way as lettering. With tones, however, it is easy to miss small areas or individual dots or pattern elements when rubbing down. To replace the sheet in exactly the right place in order to go over the missing area again can be difficult. Dry-transfer tones therefore need to be rubbed down slowly and carefully, changing the direction of the burnishing from time to time so that coverage of the selected area is complete. Windows or tracks may have to be made in toned areas to allow self-adhesive tapes or lines to stand out clearly (see Section 12.5.3).

12.4.3. Making corrections

Any unwanted or misplaced dry-transfer items can be easily and neatly removed with a piece of drafting tape, provided that they have not become too firmly fixed to the paper. The drafting tape should be carefully and precisely burnished down over the unwanted item, and then peeled off carrying the dry-transfer material with it (Figure 12.8).

12.5. Use of self-adhesive tones and colours

12.5.1. Methods of application

There are two ways of applying self-adhesive tone and colour sheets. In both cases, transillumination of the drawing is essential for speed and efficiency. (All comments made in relation to tone sheets also apply to colour sheets.)

Method 1. The tone sheet and its protective backing are placed over the drawing, and the area required is cut out precisely (Figure 12.9A). This procedure is easiest if the margins of the area to be covered by the tone are drawn in a fairly thick line (e.g. 0.7 mm). It is important to cut the tone sheet only, and not the backing. A very light touch with a sharp scalpel is all that is required. The cut-out area is then removed from the backing sheet and placed on a spare piece of backing which forms a convenient 'carrier'. The piece of toning should overlap the edge of the carrier. The toning is then placed exactly over the required area, using the carrier to guide it into

172

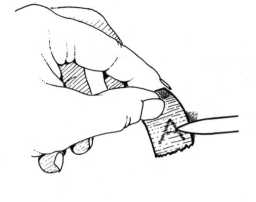

To remove unwanted dry-transfer items, simply burnish a piece of drafting tape over the letter or symbol.

Then peel the tape slowly away. The unwanted item will usually come away with the tape.

Figure 12.8   Using drafting tape to make corrections.

position. The free edge of the tone piece can then be pressed down and the carrier removed. A check on the accuracy of the positioning can then be made before the tone is finally burnished down (Figure 12.9B).

Method 2. It may be difficult to use Method 1 where very complex shapes need toning, and the following procedure should then be adopted. It is essential to use the 'low-tack' variety of material for this method. A piece of tone sheet which is more than large enough to cover the required area is first of all removed from the backing sheet. It is then placed over the drawing in the appropriate position. Any previously applied dry transfer symbols should have been well burnished down by this stage so that they will not be accidentally removed. The tone or colour is then lightly rubbed down with a finger, over the area where it is to remain. The precise area required is cut out with a sharp scalpel, taking care that only the tone sheet is cut and not the underlying paper. Finally, all excess pieces are removed, leaving the complex area neatly covered. Burnishing completes the task.

Tone sheets should never be applied over self-adhesive tapes. This not

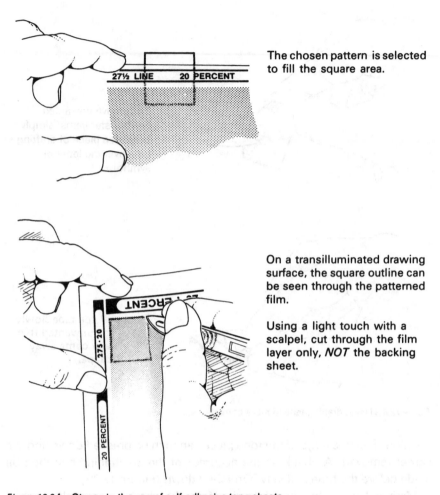

The chosen pattern is selected to fill the square area.

On a transilluminated drawing surface, the square outline can be seen through the patterned film.

Using a light touch with a scalpel, cut through the film layer only, *NOT* the backing sheet.

Figure 12.9A Stages in the use of self-adhesive tone sheets.

only causes distortion of the image, but can also create destructive highlights in photography.

12.5.2. Trapped air bubbles

The larger the area to be covered by the tone or colour, the greater is the danger of trapping air bubbles between the paper and the tone sheet. If allowed to remain, these bubbles will cause undesirable reflections during photography and will create colour differences in the case of colour sheets.

174

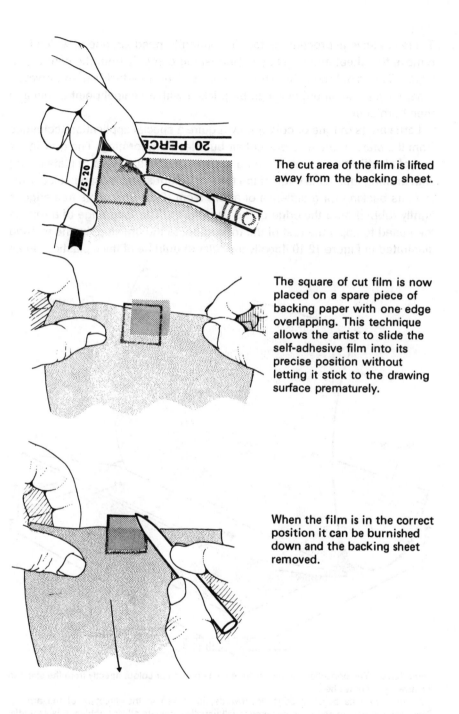

The cut area of the film is lifted away from the backing sheet.

The square of cut film is now placed on a spare piece of backing paper with one edge overlapping. This technique allows the artist to slide the self-adhesive film into its precise position without letting it stick to the drawing surface prematurely.

When the film is in the correct position it can be burnished down and the backing sheet removed.

Figure 12.9B Stages in the use of self-adhesive tone sheets continued.

There are several procedures for eliminating trapped air, one of which is to remove the sheet and re-apply it, burnishing carefully from the centre outwards. This is only possible if the material is of the 'low-tack' variety however. Alternatively, small bubbles can be pricked with a scalpel point or pin and then burnished.

Large areas of tone or colour may require a special application technique from the start, in order to prevent air bubbles from forming. The drawing is first of all affixed to a smooth, hard surface (such as a sheet of plate glass) with drafting tape. One edge of the tone or colour sheet is then peeled away from its backing for a distance of about 5 cm (2 in) only. This free edge is lightly rubbed onto the edge of the illustration. The long edge of a ruler is then used to apply the rest of the tone sheet to the drawing, by the method illustrated in Figure 12.10. Ideally the ruler should be of thick plastic with an

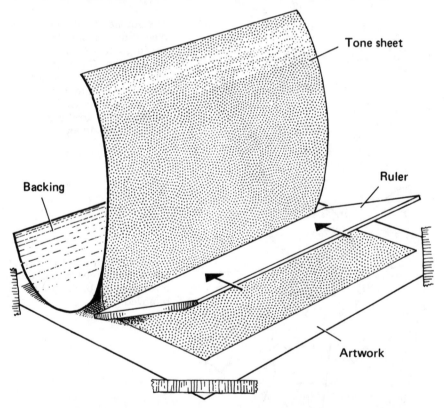

Figure 12.10 The principle of applying large areas of tone or colour directly from the sheet to the artwork is shown here.

A ruler, with its bevelled edge downwards, is moved in the direction of the arrows. Reasonable pressure and slow movement will usually eliminate all air bubbles. It is essential to work over a hard surface such as plate glass, and to secure the artwork firmly.

176

undamaged bevelled edge. It should also be longer than the width of the tone or colour sheet being applied.

12.5.3. Windows

Tones should not be applied over lettering, for the reasons already mentioned in Section 5.2.8. Most background patterns are likely to impair legibility to some extent, and some will affect it severely. Any adverse effects will be accentuated by subsequent photographic and printing processes. It will therefore be necessary to make a 'window' in the toned area to accomodate the lettering (see Figure 12.11). The edges of the window should be at least 2 mm away from the lettering on standard format artwork.

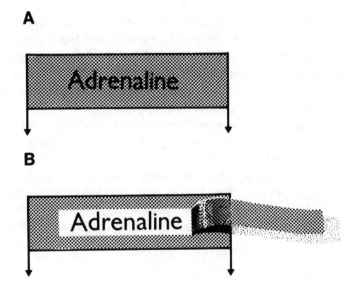

Figure 12.11 When tones are used over lettering, a rectangular 'window' should be cut and lifted off as in *B*.

Some self-adhesive tone sheets have the image printed on the upper surface of the film, and in this case it is possible to apply the tone sheet over the lettering and then scrape away the unwanted area of the image. Sometimes, however, the tone is printed on the lower surface of the film, and it is then necessary to cut a window.

This can be done either after the tone sheet has been applied but before it is finally burnished down, or when the backing sheet is still in place. In the latter case, for the sake of accuracy, it is essential to fix the tone sheet (plus backing) in position over the drawing by attaching it to the drawing board with drafting tape. This eliminates any slight risk of movement during the cutting operation. This second method is recommended where dry transfer symbols are already in position in a drawing, as there will then be no danger of accidental removal of the symbols.

Windows should be cut very carefully with a sharp scalpel. A ruler and a set square should be used to obtain sharp corners and true right angles where rectangular openings are required.

12.6. Simple object drawing for line diagrams

12.6.1. General principles

The art of object drawing is based on the understanding of only four basic shapes, and this combined with a very simple approach to lighting can give added scope to those who wish to develop their illustrating powers. An uninhibited desire to try something new, together with a certain amount of straightforward observation will lead the interested amateur to the point where his work will be of publishable quality.

The four most useful shapes in three dimensional drawing are illustrated in Figure 12.12. The representation of light and shade is necessary to give a three-dimensional effect.

Basic lighting is that which is imagined as coming from the top left hand edge of the page. Thus the most brightly lit areas of a drawn object will be the top and the left hand side as seen by the observer. Conversely, the areas of strongest shadow will be the right hand side of the object and its lower edge.

12.6.2. Basic shapes

A rectangle can be made to suggest a solid object if its dark side is emphasised, as shown in Figure 12.13. The use of an oval template will change the rectangle into a cylinder, as in Figure 12.13. A 30° oval will be suitable for most purposes, but 20° and 40° ovals are also acceptable (Figure 12.14). The more circular the oval, the greater is the imagined elevation of the observer's eye. In other words, more of the top of the object is visible and less of the sides. A cylindrical vessel can be made to appear empty or full of liquid by using the same template in the manner shown in Figure 12.13. The effect is enhanced by the use of light and shadow. Too much realism should be avoided however, as it may destroy the effect of a simple diagram.

178

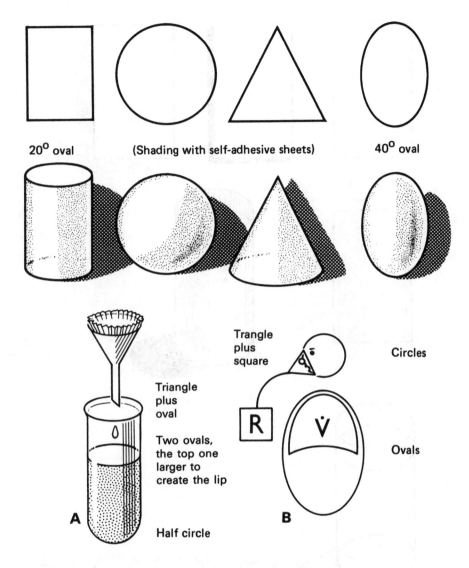

Figure 12.12 Four basic shapes as flat line drawings, as solids and as used to describe processes. These shapes are all available in different sizes as cut-outs in plastic templates. Diagram *A* took 16 minutes to draw, and diagram *B* took 8 minutes.

A square can be used to draw cubic shapes by the method shown in Figure 12.15. With a 45° + 45° + 90° or a 30° + 30° + 90° set square, all the angles necessary for representing a cube can be drawn. Turning the original square through 30° or 45° and making the vertical side equal to the other sides of the square creates isometric perspective. This is useful be-

179

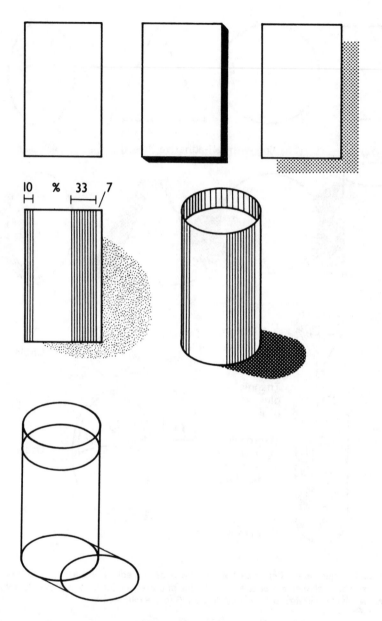

Figure 12.13 Ways of giving a rectangle different degrees of solidity, and some steps towards turning it into a cylinder filled with liquid. Note the same oval (35°) is used throughout. All shading effects are achieved by using different sorts of self-adhesive tone sheets.

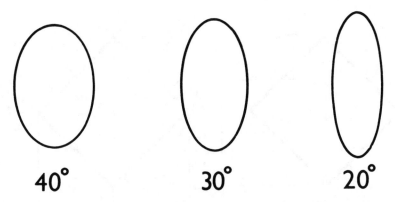

40°   30°   20°

Figure 12.14 Three of the most useful angles for ellipses in diagram drawing. Templates are available which provide each angle in many different sizes.

cause all measurements along any side of the square will then be true measurements, and not disturbed as they would be in the case of optical (aerial) perspective. (In the case of optical perspective, distant objects appear smaller. This is not so with isometric perspective.)

12.6.3. Lighting

If shadows look convincing, then the belief in a light source is created. The appropriate distribution of shadow on a drawing is therefore very important. As a rule, cast shadows should be the darkest areas of all on a drawing.

The cylinder and sphere demonstrate an important property of all curved surfaces with respect to shadows (Figure 12.16). The area in shade does not extend to the edge of the dark side of the object. A narrow band is left white or only lightly shaded. This is because the shaded area is receiving reflected light from other surfaces. If the sphere is lifted away from the flat surface, the cast shadow becomes softer at its edges.

12.6.4. Special effects

Sometimes it is necessary to create an impression of transparency, either for purely diagramatic reasons, or to give the effect of glass or clear plastic material. This can be achieved by scraping the drawing so that some lines are partially destroyed. The technique is illustrated in Figure 12.12.

Differences in line thickness can be used to suggest simple perspective.

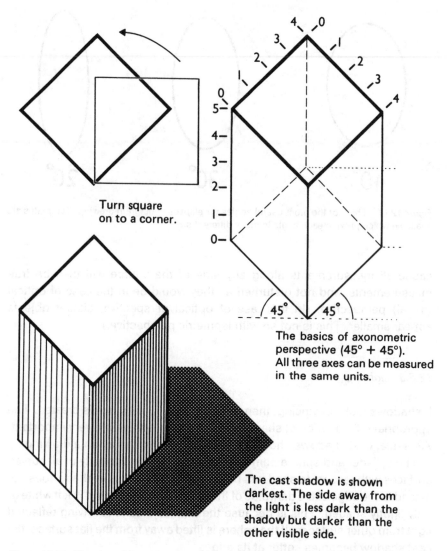

Turn square on to a corner.

The basics of axonometric perspective (45° + 45°). All three axes can be measured in the same units.

The cast shadow is shown darkest. The side away from the light is less dark than the shadow but darker than the other visible side.

Figure 12.15 The use of axonometric techniques. Note that light is always assumed to be coming from the top left.

Thick lines appear to be nearer to the observer than thin lines, and those parts of the object which are closest to the observer can therefore be stressed in this way with good effect.

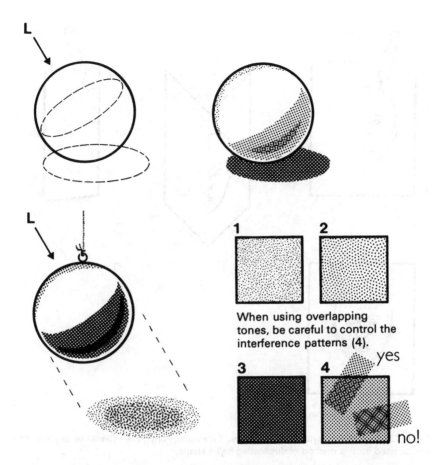

Figure 12.16 Notes about light and shade. Guide lines are shown as thin dashed lines and these are drawn on the *back* of the paper so that when the sheet is transilluminated they show where the tone sheets should be cut. When an object lies on a surface, its cast shadow tends to be darker. When it is suspended above a surface, the cast shadow gets lighter with distance; sometimes this shadow may have a darker centre.

12.6.5. *Making symmetrical shapes*

Many symmetrical shapes can be drawn by the following method: draw only half the design on thin typewriter copy paper then fold this in half. A clear image of the design already drawn will be seen, now laterally reversed; this can then be drawn on the second half of the sheet. When unfolded the drawing will be symmetrical (see Figure 12.17).

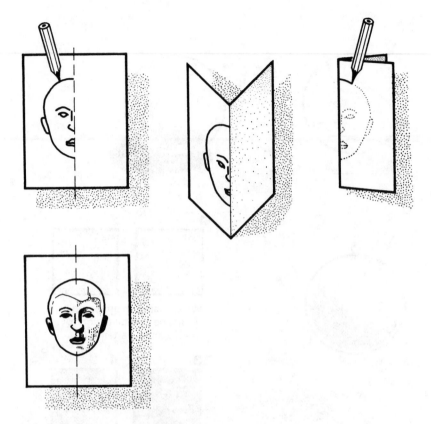

Figure 12.17 Drawing symmetrical shapes. Typewriter copy paper (bank) or any thin paper can be used for this method of duplicating half a shape.

12.6.6. Observation of objects

The appearance of any real object is made complex by the effects of detail and of light and shade. These distractions can be reduced by looking at the object through partly closed eyes. This will almost eliminate the effects of colour and will blur unimportant detail so that the basic shape will be more readily perceived. The illustrator has then to ask only four questions: Is it a rectangle (cylinder), a square (cube), a triangle (cone) or a circle (sphere)?

184

13. Working method

13.1. Planning the illustration

13.1.1. Choice of media

Before an illustration can be planned in detail, a decision must be made as to the medium or media for which it is intended. The various media discussed in this book (i.e. printed publications, slides, posters, etc.) each have their own requirements and limitations. The best illustrations are those designed for presentation in one medium only, but in most cases it will be necessary to use the same artwork for more than one purpose, for the sake of economy. With careful planning, however, it is often possible to reach a satisfactory compromise. If the same artwork can be used for publication, for slides and for posters, then a difficult ideal has been achieved.

The significant characteristics of the various media in which an illustration may be presented can be summarised as follows:

Publication in books. Books tend to be regarded as a 'permanent' record, and the illustrations must therefore be of exceptional quality. For this reason, the publisher will often undertake responsibility for the artwork. Where the author prepares the artwork, special care must be taken to fit the illustration to the space available, and to provide lettering as specified by the printer. Several colours may be used in book illustrations, and this may restrict the use of artwork for other media (see Chapter 5).

Publication in journals. Rapid execution of the artwork with minimal sacrifice of quality is important here. Illustrations will usually be in black and white, and will be designed to fit across one or two columns on the printed page. The shape of such illustrations may be unsuitable for use in other media unless this requirement is borne in mind from the beginning. Lettering may or may not be required on journal illustrations. If lettering is not required and the artwork is also to be used to make a slide, for example, then it will be necessary to make high quality copies of the artwork so that lettering can be inserted subsequently (see Chapter 5).

Slides. It is essential that slides should be legible from the *back* of the lecture theatre, and this requirement will determine the minimum capital letter height, the stroke thickness of the letter forms, and the minimum line thickness. Furthermore, artwork for slides must conform with standard slide formats if the available image area is to be used efficiently. These requirements may conflict with those for other media (see Chapter 6).

Posters. The average viewing distance for poster displays is between 60 and 90 cm (2-3 ft), and illustrations must therefore be legible from this distance. There are few limitations on illustration format however, and artwork intended for use in some other medium may therefore be suitable (see Chapter 7).

Overhead projector transparencies. The usual format for OHP transparencies is 25 × 25 cm (10 × 10 in). These displays are generally viewed at relatively close quarters, and this will influence the choice of size of lettering and line thicknesses. As a result of these characteristics, illustrations prepared as OHP transparencies are rarely suitable for use in other media (see Chapter 8).

Tape-slide illustrations. Tape-slide programmes are viewed at a normal reading distance, and the illustrations will therefore have more in common with those intended for books and journals than with those suitable for lecture slides. The format of the illustrations may also differ from conventional slide formats, depending on the equipment in use. Tape-slide illustrations may not be suitable for any other purpose therefore (see Chapter 9).

Television illustrations. These require a special design approach, and will rarely be suitable for use in other media (see Chapter 10).

Temporary or one-off illustrations. Illustrations are often required at very short notice. If these are to perform efficiently, the formats and minimum sizes of lettering must conform with the standards suggested in this book for the medium concerned. Provided that these requirements are met, the illustrations will be legible even if executed on a casual basis.

Multi-media illustrations. Many scientific illustrations need to meet the requirements for at least two different media. The most common situation is where the same artwork must be used for publication in a journal and for the preparation of a lecture slide. With careful planning a successful compromise can usually be achieved. It may also be possible to use the same artwork for a poster display.

13.1.2. Summary of standards

Wherever possible, artwork for illustrations should be prepared within an

A4 format. This will make for convenience in handling, and will also facilitate the use of the same piece of artwork for more than one presentation medium.

The image area on the artwork should be approximately 200 × 130 mm (8 × 5 in). In the case of illustrations for publication a vertical format is usually preferred, but either format may be used for projection slides.

The capital letter height of the lettering chosen should not be less than 1/40th of the longest dimension of the image area. Thus a drawing measuring 200 mm on its longest dimension requires a capital letter height of 5 mm.

In the case of printed publications, it is the horizontal dimension of the image area which is important, since the illustration must fit within a fixed column width. Assuming a 50% reduction ratio, artwork for an illustration intended to fit within a 75 mm (3 in) column width should have a horizontal dimension of 130 mm. Given that the minimum acceptable capital letter height for the printed illustration will be 2 mm, the minimum capital letter height required on the artwork will be 4 mm.

For artwork with an image area of 200 × 130 mm up to full A4 size, the minimum acceptable line thickness will be 0.35 mm and the maximum will be 1.0 mm.

13.2. Making the draft

13.2.1. Rough sketches

Having decided on the size and format of the artwork, it will then be possible to make a rough sketch of the illustration to determine the relative positions of the various parts of the drawing and of any captions and labels. An accurate draft should then be made on the basis of the rough sketch.

13.2.2. The use of graph paper for drafts

It is often convenient to make the accurate draft on graph paper, particularly if the illustration is a graph or chart. The various elements of the illustration, including both lines and lettering, should be carefully positioned so that their relationship to one another is clearly shown, and so that no part of the drawing is cramped or unduly spaced out. Lines should be drawn boldly and accurately, and the exact amount of space which will be occupied by lettering in the appropriate size should be clearly indicated on the draft. The positioning of lettering will be facilitated if the size of the grid squares on the paper bears a simple relationship to the letter size.

13.2.3. *The use of typewriter grids*

When the lettering on an illustration is to be typewritten, it is often helpful to use a typewriter grid when preparing the draft. Such a grid can easily be prepared by typing lines of full stops over an area measuring 200×130 mm on an A4 page. Each full stop then represents one typewritten character position. This grid should be fixed to the back of the draft or to the drawing board, as described in Section 13.3 below. On a transilluminated drawing surface, the grid can then be used to make an accurate draft of both line work and lettering. The scales for graphs and charts can be based on the typewriter character units, which will simplify the positioning of scale point markers and labels. The accurate determination of the position of all lettering in this way will ultimately mean that the artwork, when placed in the typewriter, will not require constant repositioning. Such a grid is also invaluable in the planning of typewritten tables. However, this method does not work with typewriters using proportional spacing.

13.3. Using the draft

13.3.1. *Fixing the draft to the artwork or drawing board*

There are two ways of fixing drafts so that they can be accurately reproduced on good quality paper. The draft may be fixed to the back of the final artwork, as shown in Figure 13.1B. In this case, the artwork and draft may be moved at will to ensure that the artist's hand is always in a comfortable position. This system is particularly useful for illustrations where curves or freehand drawing are required. Alternatively, the draft may be fixed directly to the drawing board, as shown in Figure 13.1A. The artwork is then also fixed to the drawing board, but independently of the draft. This system is preferable where rulers, set squares, templates or stencils are in use.

13.3.2. *Transferring information from the draft to the final artwork*

A transilliminated drawing surface is essential. Transillumination imparts a 'see-through' quality to any good drawing paper. By this means all the sketch lines, lettering etc. on the draft are made visible and can be copied directly onto the final artwork.

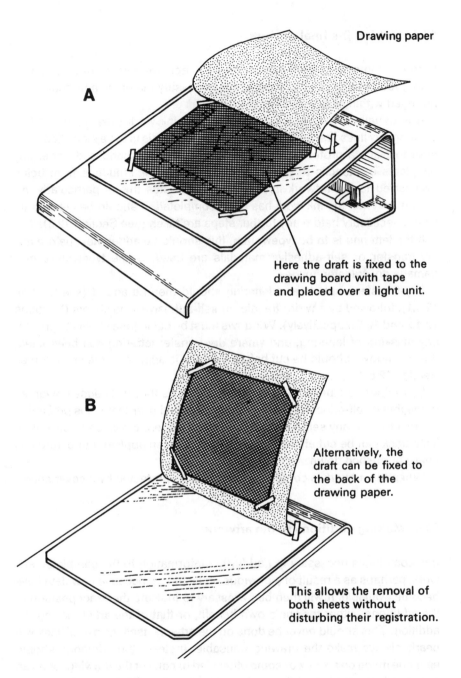

Drawing paper

A

Here the draft is fixed to the
drawing board with tape
and placed over a light unit.

B

Alternatively, the
draft can be fixed to
the back of the
drawing paper.

This allows the removal of
both sheets without
disturbing their registration.

Figure 13.1   Using a draft.

13.4. Making the final artwork

Before commencing work it is important to achieve a comfortable position with good lighting, and to have all the necessary materials and equipment arranged within easy reach.

It is usually convenient to begin with the ink work (see Section 12.2), drawing any lines first and then adding lettering if this is to be stencilled. (It is wise to place a protective sheet of paper under the hand and under any templates or stencils to prevent perspiration or ink smudges from being transferred to the artwork itself. In warm climates this is particularly important.) When the ink work has been completed, it should be cleaned up where necessary before any further steps are taken (see Section 12.3).

If the lettering is to be typewritten, this should be added next before any dry-transfer or self-adhesive materials are used. Some typewriters may damage dry-transfer images.

Dry-transfer symbols and lettering should then be added (see Section 12.4.1), followed by any dry-transfer or self-adhesive tone sheets (Sections 12.4.2 and 12.5 respectively). Windows must be cut in tone sheets to prevent any masking of lettering, and where dry-transfer lettering has been used, these windows should be cut before the tone is applied to the artwork (see Section 12.5.3).

Self-adhesive tapes should be applied last, as they will cause unwanted highlights if self-adhesive tone sheets are applied over them. It is preferable to cut a track in any self-adhesive tone sheets where tapes are to be added. This track can be cut after the tone sheet has been applied, but before it is finally burnished down.

When the artwork is completed, it should be protected by a cover sheet.

13.5. Making additions to the artwork

It is sometimes necessary to add further information to the completed artwork, perhaps as a result of a previous omission, or because new data have become available. It will then be necessary to indicate the exact position of the additions, either for one's own benefit, or that of the artist making the addition. This should never be done on the artwork itself, as pencil lines will nearly always make the drawing unusable. Instead, the additions should either be made on a xerox or some other kind of copy of the drawing, or on an overlay attached to the original with drafting tape (see Section 5.4.). There is, however, a disadvantage in using any copying process such as xeroxing in which heat plays an essential part. Heat distorts paper, and usually only in one direction. This distortion can cause serious errors if extreme accuracy is required. Depending on the quality of the paper, an A4 sheet may shrink as much as 5 mm when xeroxed.

When the position of the additions has been indicated in one of these ways, they can be transferred to the artwork by placing the copy or overlay beneath it on a transilluminated drawing surface.

Alternatively, it may be convenient to use the following method instead. The drawing is placed face downwards on the transilluminated surface, thus showing it in full detail though laterally reversed. The additions are then indicated on the back using a soft 2B pencil. This will make a line sufficiently dense to be seen when the drawing is placed face upwards. In this position, the new material can be drawn in using the guide lines on the back of the drawing. This system is particularly useful when adding arrows or other marks to photographs. The required position of these will be easier to see if they are marked on the back of the photograph than if they are on a draft beneath it.

Glossary – some terms used in the graphic arts and in statistics*

Abscissa: The horizontal or 'x' axis of a statistical grid.

Acetate, *Foil, Cell*: Clear film used for overlay work.

Align: To arrange letters, words etc on the same vertical axis.

Annotation: An explanatory note forming part of an illustration.

Arabic numerals: The figures in general use 1 2 3 4 5 6 7 8 9 0, as distinct from Roman numerals.

Arithmetic mean: The sum of the values recorded in a series of observations divided by the number of observations.

Art board: Board with a high quality finish for preparation of drawn artwork.

Art paper: Paper which is coated, usually with china clay, to give a very smooth glossy surface, suitable for printing half-tone blocks.

Artwork, *a/w*: Also called original or master. The final illustration and/or type in a form suitable for presentation to the printer or photographer. (a/w may include photographs prepared for the printer).

Ascender: The stroke which extends above the main body of the character as in 'h'.

Asymmetry: Unequal proportion. Uneven distribution of data.

Average: A measure of the most 'typical' value in a series of observations. There are three ways of expressing averages: arithmetic mean, median or mode.

Axis: A fixed line adopted for reference. A line, about which a body rotates, or about which the parts of a figure etc. are arranged. Graphs or charts are usually organised on axes which are at right angles to each other.

Balanced layout: A layout which is 'balanced' by eye and not by measurement. A balanced layout is not necessarily symmetrical.

Base line: The imaginary line on which the data stand, e.g. the zero line in a bar chart or histogram.

Bleed (printing): A term used to describe an illustration which extends to the trimmed edges of the paper.

Block: Used in letterpress printing, for line illustrations and/or half-tone work.

Blow-up (printing): An enlargement.

Body (type) (texts): 1. Rectangular piece of metal which carries the letter or character for letterpress printing. 2. The main text of a book, report or paper.

Bold face (type): Thicker than normal type.

Brief: Instructions received from a client.

Bristol board: Top quality rag paste board with a very smooth finish.

Burnish: To rub with a special tool, for example to smooth a surface or to ensure proper adhesion of dry-transfer products.

* Italic type = synonyms. ( ) = relevant subject.

Calibrations: The divisions of a scale.

Camera-ready copy: Text and/or illustrations ready for immediate reproduction in print.

Caption: An identifying label.

Cartridge paper: A strong, rough-surfaced drawing paper.

Cell, *Acetate, Foil*: A sheet of transparent film, either clear or coloured.

Centred type: Type which, when set, is aligned so that the mid-point of each line is directly under the mid-point of the previous line.

Character: The individual letter, numeral or punctuation mark. Note: when counting characters and calculating the space they will occupy, it is essential to count inter-word spaces as one character.

Coaxial: Lying on the same axis.

Column width: The width of one column of type as in books and journals.

Colour guide: A print or sketch of an original indicating colours to be used by the printer (see colour swatch).

Colour separation: When an illustration is to be reproduced in several colours, the artist is often required to make a separate drawing in black and white for each colour involved. Black-and-white original 'separations' are necessary for making separate printing plates. The separations must register accurately with each other and the basic artwork.

Colour swatch: Reference colours supplied by manufacturers of ink, paints, paper, etc.

Compose: A printer's term for setting type. Carried out by a compositor.

Condensed face (type): A style of type which is narrow in width.

Contact print: A print the same size as its negative.

Continuous-tone: See Half-tone.

Co-ordinate: A precise reference, which locates a point, line or plane, in 2 or 3 dimensional space.

Copy: The manuscript or sketches containing the information to be reproduced. Usually refers only to text or captions.

Counter (type): The space inside certain characters; a, o, e are examples of letters with counters.

Cover sheet, *Overlay*: Protective cover to keep illustration clean.

Cross hatching: The criss-cross patterns made with pen and ink to simulate textures.

Cut-away view: An illustration showing an object partly sectioned to reveal internal details.

Dependent variable: A variable which is altered by changes in the independent variable. Dependent variables should always be placed on the vertical axis of a graph.

Descender: The part of the type which extends below the main body of the character, e.g. 'g'.

Diazo: Slide with white illustration on a coloured (often blue) background.

Dimensionally stable: Material which will not distort, shrink or stretch. Important where print registration is essential.

Dot for Dot: A method of same size reproduction of half-tones, used in screen printing, which ensures that the screen images coincide.

Double (page) **spread**: Both left and right hand pages of a book or journal, or the layout for the same.

Draft (rough): The manuscript sketch, providing the basic information necessary to create an original or final illustration.

Dry mounting: A method of fixing card, paper, photographs, etc., using adhesive either in the form of a shellac sheet which becomes sticky when heated, or in the form of wax from an aerosol can.

Dry-transfer: A general term for 'rub down' lettering or symbols.

Dye transfer: A process of print-making using photo mechanical transfer (PMT) of dyes either in colour or black-and-white.

Elevation: A scale drawing of objects viewed from the side.

Expanded (type): A style of type in which the horizontal width has been increased in relation to the vertical height of the letter.

Explicit table: A complete table in which all values are explicitly stated.

Exploded view: An illustration showing an object partly or wholly dismantled to show the components in order of assembly.

Face (type): The part of the type or plate which makes contact with the paper.

Flush: See Range.

Font: A complete family of characters of one size and style of type.

Format: The shape of the illustration. See Horizontal format and Vertical format.

Frame: The outer limits of an illustration, including any necessary spaces or margins.

Frequency: If a set of data is divided into categories, the number of items in each category is known as the frequency distribution.

Glossy prints, *Glossies*: Photographic prints with a shiny surface.

Gouache: Opaque water colour.

Grey scale: See Half-tone.

Grid: A framework (giving type areas, margins, illustrations, etc.) used as a guide for laying down elements within a given format.

Hairline spacing: Very fine spacing between individual letters or characters in a text.

Half-tone, *Continuous tone, Grey scale*: A printing technique in which continuous tones (greys) are broken down into dots.

Hierarchical order: Order of importance.

Highlight: The lightest tonal value of a half-tone.

Horizontal format, *Landscape*: The shape of an illustration when the horizontal measurement is greater than the vertical; also known as landscape or comic mode.

Hue: Colour of light measured by wavelength, but given inaccurate, often subjective names which cannot be used in correspondence with printers and others (e.g. 'pink' or 'brown').

Implicit table: A summary table giving basic information from which specific values must be calculated by the user.

Independant variable: Any variable whose values are *not* affected by changes in other variables. Time is an example. Independent variables should always be placed on the horizontal axis of a graph.

Italic, *ital.*: A sloping style of character.

Justified type: Type which when set has both left and right margins aligned.

Key: Inset information describing parts of an illustration.

Lantern slide: A photographic transparency designed for projection onto a screen. This term is archaic and inaccurate and should be replaced by using the words – 'projection slide' so as not to confuse it with microscope slide.

Layout: A design in the form of a pencilled rough which indicates where the various parts of the artwork or design should be placed.

Leaders: Lines, dashes or dots used to guide the eye across the page. Often used to link words and the items they describe.

Legibility: Can you read it?

Letterpress: Printing by means of raised metal type.

Letter-spacing: The space between each letter or character.

Light (type): Lettering or lines using thin strokes.

Line block: A printing plate on which the image has been reproduced as a raised surface. The raised areas receive ink and thus impart the image to the paper.

Line spacing: The space in between each horizontal line of type.

Line work, *Line drawing*: The use of black-and-white high-contrast images, without using continuous tones.

Lithography: A grease/water resist method of printing.

Logarithmic scale: Scales which are subdivided according to logarithmic principles. These allow large variation in quantities to be shown on the same illustration. Log scales always start at 1.

Lower case, *Uncials*: All letters which are not capitals.

Mask: An opaque cut-out overlay used to mask out unwanted portions of an illustration.

Master: See Original.

Matt print: A photographic print with a dull finish.

Mean deviation: The arithmetic average of all the differences between the observations and their mean.

Measure: The width of the type area, or line length. Usually measured in centimetres or inches, but sometimes in picas.

Median: The centre value of a series of observations (a form of average).

Microfiche: A flat rectangular sheet of photographic film, usually 105 mm × 148 mm, containing multiple micro-images of text or diagrams in a grid pattern of rows and columns. A microfilm display unit (reader) is necessary for reading purposes.

Micronorm: An ISO (International Standards Organization) standard for draftsmen which controls use of letter forms and line thickness when preparing artwork for microfilming.

a) The stroke width of lettering is 1/10 of the letter height.

b) The differences between letter heights are controlled by the proportion of 10:7 which gives a multiplication factor of 1.4. All lettering sizes are determined from a starting point of 10 mm.

c) The rules also apply to line thicknesses and the standard range consists of nine line widths as follows: 0.13, 0.18, 0.25, *0.35, 0.5, 0.7, 1.0,* 1.4 and 2.0 mm. The italicised series are recommended for the 13 cm × 20 cm artwork formats mentioned in this book.

Mirror image: Lateral reversal so that left on the original becomes right on the copy.

Mode: The most frequently occurring value in a distribution.

Monochrome: In one colour only.

Montage: The fitting together of several images to create a new picture.

Negative slide: A slide on which the tones of artwork or objects are reproduced as their opposites, i.e. black is seen as white.

Opaque: Does not transmit light.

Optical (*aerial*) **perspective**: A form of drawing which seeks to simulate form as seen by the human eye.

Ordinate:The vertical or 'Y' axis of a graph or chart. Can also be any vertical line which bisects the abscissa.

Orientation: The direction of the longest dimension of an object or illustration.

Original: The final artwork from which copies can be made.

Overlay: A translucent cover fixed to an illustration, often used to pencil in instructions to printer, photographer, editor, etc.

Pan: TV or cine camera movement which traverses the artwork either horizontally or vertically.

Paste-up: An illustration consisting of various parts pasted together.

Percentiles: If the data is divided into 100 equal parts, each part is called a 'percentile'.

Perspective: A three-dimensional view.

Phototype: Type that is set and reproduced photographically.

Pica: Twelve 'points', or approximately 4 mm (1/6 in). Sometimes used in specifying line length.

Pitch (typewriters): The number of characters to 2.5 cm (1 in).

Plan: A drawing of objects viewed from above.

Point size: A printer's measure for type. 72 point = 2.5 cm (1 in).

Positive slide: A slide which reproduces the tones of artwork or objects as they are in the originals (i.e. black is seen as black).

Proofing paper: High quality coated paper used by printers to make proofs of illustrations.

Proofs: These are printed sheets of text or illustration in their intended final form. They are sometimes supplied to artist or author for checking before the main print run is started.

Quartile: The centre point on a line, equidistant between the median and the extremes.

Range: The difference between the lowest and highest values observed.

Ranged left, *Flush left*: Type which is aligned on the left hand side, whilst the right hand side is ragged.

Ranged right, *Flush right*: Opposite of 'ranged left'.

Readability: Does it make good sense quickly?

Ream: 500 sheets of paper.

Rectilinear, *Rectangle*: A shape composed of lines at right angles to each other.

Recto: A right-hand page of a publication. Opposite of 'verso'.

Reduction ratio: Ratio of the size of original illustration to its final size – properly expressed as a percentage (1:1 = 100%).

Reflectance: The brightness of reflected light, often depending on the nature of the reflecting surface, ie its colour, texture etc.

Register marks: Corresponding marks on an original illustration and its overlays to ensure correct alignment. Usually indicated by fine crosses outside the reproduction area.

Reprography: The complete process of reproducing artwork on to a flat surface.

Reversal (photography): White image from black original or vice versa.

Reversed lettering: This has two meanings and care must be taken in using these words:

    a) Lettering which has been laterally reversed, i.e. mirror writing.

b) Lettering which is reversed to its opposite tone. Thus black lettering is reversed to white or vice versa.

Roman, *Regular*: An upright (vertical) style of lettering.

Same size: Identical size as original. Sometimes abbreviated to s/s.

Sans serif: A typeface without serifs (see Serif).

Saturation: Purity of a colour – closeness to one of the six basic colours, ie red, orange, yellow, green, blue, violet.

Serif: A finishing or decorative stroke to the ends of lettering. (Highly developed in Roman stone cut letter forms).

Shading tint: A line or dot pattern that can be added to an original to simulate tonal values.

Standard error of the mean: The square root of the arithmetic mean of the squares of the differences between the observations and their mean, plus one.

Stipple: A dotted texture.

Symmetry: Exact correspondence in size and shape between opposite sides of a figure.

Tabulation: The systematic arrangement of data into columns.

Texture: A tone or tint applied to an illustration by drawing dots, lines, cross-hatching etc. Applied with skill, textures can be made to simulate tone, leather, wood, glass, metal, etc.

Track: TV or cine camera movement towards or away from artwork.

Transparency: An ambiguous word for projection slide, also meaning overhead projection acetate.

Trim: The limits to which an illustration may be trimmed, as indicated by corner marks.

Type size: Size of type, as measured by its capital letter height or the depth of its metal body.

Typography: Art of designing with type or typewriting.

Underline, *Underscore*: A line placed below lettering or symbols.

Upper case: Term for a capital letter.

Variable: Data subject to measured change.

Variance: The square of the standard error of the mean.

Verso: A left-hand page of a publication.

Vertical format, *Portrait*: The shape of an illustration when the vertical measurement is greater than the horizontal.

Visibility: Can you see it?

Weighted average: The arithmetic average is multiplied by a suitable 'weight' corresponding to importance.

X-height: The height of a lower case letter form, not including ascenders or descenders.

Literature

Source works

British Standards Institution (1976) *Bibliographic references: recommendations*. BS 1629 London: BSI.

Burnhill P, Hartley J and Davies L (1977) *Typographic decision making: the layout of indexes*. Applied Ergonomics 1977 *8* (1) 35-39.

Burnhill P, Hartley J and Young M (1976) *Tables in text*. Applied Ergonomics 1976 *7* (1) 13-18.

Burt CL (1959) *A psychological study of typography*. Cambridge: Cambridge University Press.

Burt CL, Cooper WF and Martin JL (1955) *A psychological study of typography*. British Journal of Statistical Psychology 1955 *8* 29-57.

Carter LF (1947) *An experiment on the design of tables and graphs used for presenting numerical data*. Journal of Applied Psychology 1947 *31* 640-650.

Cattell JMcK (1885) *Über die Zeit der Erkennung und Benennung von Schriftzeichen, Bildern und Farben*. Philosophische Studien 1885 *2* 635-650.

Dooley RP and Harkins LE (1970) *Function and attention-getting effects of colour on graphic communications*. Perceptual and Motor Skills 1970 *31* 851-865.

Dwyer FM (1976) *The effect of IQ level on the instructional effectiveness of black and white and colour illustrations*. Audio Visual Communications Review 1976 *24* 49-62.

Erdmann B and Dodge R (1898) *Psychologische Untersuchungen über das Lesen, auf experimenteller Grundlage*. Halle: Max Niemeyer.

Frase LT (1969) *Tabular and diagrammatic presentation of verbal materials*. Perceptual and Motor Skills 1969 *29* (1) 320-322.

Galer IAR (1976) *Projector slides – preparation, construction and use*. Applied Ergonomics 1976 *7* (4) 190-196.

Goldscheider A and Müller RF (1893) *Zur Physiologie und Pathologie des Lesens*. Zeitschrift für Klinische Medizin 1893 *23* 131.

Gregory M and Poulton EC (1970) *Even versus uneven right-hand margins and the rate of comprehension in reading*. Ergonomics 1970 *13* 427-434.

Hailstone M (1973) *A case for standardisation in the preparation of graphs and diagrams*. Medical and Biological Illustration 1973 *23* 8-12.

Hartley J (1978) *Designing instructional text*. London: Kogan Page.

Hartley J, Young M and Burnhill P (1975) *On the typing of tables*. Applied Ergonomics 1975 *6* (1) 39-42.

Javal E (1905) *Physiologie de la lecture et de l'ecriture*. Paris: Felix Alcan.

McCormick EJ (1976) *Human factors in engineering and design*. New York: McGraw Hill Co.

Messmer O (1903) *Zur Psychologie des Lesens bei Kindern und Erwachsenen*. Archiv für die gesamte Psychologie 1903 *2* 190-298.

Morton R (1968) *The lantern slide: Legibility and production*. The Photographic Journal 1968 *April* 89-97.

Operbeck H (1970) *Effect of paper and ink gloss on legibility*. Journal of Typographic Research 1970 *4* (2) 187-188.

Poulton EC (1959) *Effects of printing types and formats on the comprehension of scientific journals*. Nature 1959 *184* 1824.

Poulton EC (1969) *Skimming lists of food ingredients printed in different brightness contrasts*. Journal of Applied Psychology 1969 *53* (6) 498-500.

Powers SP (1962) *The effect of three typesetting styles on the speed of reading newspaper content*. MA Thesis, School of Journalism and Communications, University of Florida.

Rawlinson G (1975) *How do we recognize words?* New Behaviour 1975 *August 28* 336-338.

Schutz HG (1961a) *An evaluation of methods for presentation of graphic multiple trends*. Human Factors 1961 *3* 108-119.

Schutz HG (1961b) *An evaluation of formats for graphic trend displays - Experiment 2*. Human Factors 1961 *3* 99-107.

Smith JM and McCombs ME (1971) *The graphics of prose*. Journalism Quarterly 1971 *48* 134-136.

Snowberg RL (1973) *Bases for the selection of background colours for transparencies*. Audio Visual Communication Review 1973 *21* 191-207.

Spencer H (1968) *The visible word*. London: Lund Humphries.

Spencer H, Reynolds L and Coe B (1975) *Spatial and typographic coding in printed bibliographical materials*. Journal of Documentation 1975 *31* (2) 59-70.

Spencer H, Reynolds L and Coe B (1977a) *The effects of show-through on the legibility of printed text*. London: Royal College of Art, Readability of Print Research Unit.

Spencer H, Reynolds L and Coe B (1977b) *The effects of image degradation and background noise on the legibility of text and numerals in four different typefaces*. London: Royal College of Art, Readability of Print Research Unit 1975, revised.

Spencer H and Shaw A (1969) *Letter spacing and legibility*. British Printer 1969 *84* 84-86.

Tinker MA (1963) *Legibility of print*. Ames: Iowa State University Press.

Tinker MA (1965) *Bases for effective reading*. Minneapolis: University of Minnesota Press.

Wilkinson GL (1976) *Projection variables and performance*. Audio Visual Communication Review 1976 *24* 413-436.

Woodward RM (1972) *Proximity and direction of arrangement in numeric displays*. Human Factors 1972 *14* 337-343.

Wright P (1977) *Presenting technical information: a survey of research findings*. Instructional Science 1977 *6* 93-134.

Wright P and Fox K (1970) *Presenting information in tables*. Applied Ergonomics 1970 *1* 234-242.

Zachrisson B (1965) *Studies in legibility of printed text*. Stockholm: Almqvist and Wiksell.

Useful reference works

Information leaflets from Kodak:*

S3	Audio visual projection
S11	Audio visual planning equipment
S13	Materials for visual presentation.
S16	Kodak projection calculator and seating guide.
S22	Effective lecture slides.
S24	Legibility - Artwork to screen.
H42	Television graphics production template.
V1-15	Slides with a purpose - Index to Kodak information.
S30	Planning and producing slide programmes.

Basic typography, by J R Biggs. London: Faber and Faber 1968

Finer points in the spacing and arrangement of type, by G Dowding. London: Benn Brothers 1974

Designing instructional text, by J Hartley. London: Kogan Page 1978

Design in business printing, by H Spencer. London: Sylvan Press 1952

Diagrams: a visual survey of graphs, maps, charts and diagrams for the graphic designer, by A Lockwood. London: Studio Vista 1969 (now out of print but available in some libraries)

Recommendations for the presentation of tables, graphs and charts (DD.52. 1977). London: British Standards Institute 1977

Charts and graphs, ed by D Simmonds. Lancaster: MTP 1980

Principles of medical statistics, by B Hill. London: Lancet Ltd 1969

Statistics in small doses, by W M Castle. Edinburgh: Churchill Livingstone 1977

Statistics, by W M Harper. Plymouth: Macdonald Evans 1976

Graphics master II, by D P Lem. USA: Dean Lem Associates 1977

Model guidelines for the preparation of camera ready typescripts by authors/typists, ed by M O'Connor, London IFSEA/CIBA 1980

Writing scientific papers in English, by M O'Connor and F P Woodford. Amsterdam: Elsevier 1975

Editing scientific books and journals, by M O'Connor. London: Pitman Medical Ltd 1978

Style for print and proof correcting, by R A Hewitt. London: Blandford Press 1957

Designing a tape slide programme, by T Graves and V Graves. Medical Education 1979 *13* 137-143

* Enquiries to: Publication Department, Kodak Ltd, Kodak House, Hemel Hampstead, Herts, UK.

Graphic design in educational television, by B Clark. London: Lund Humphries 1974

Guide to the correct selection and use of material for technical drawing. Nürnberg: Staedtler GMbH 1977

Rendering with pen and ink, by W Gill. London: Thames and Hudson 1973

Subject index